D1194755

# DNA ISOLATION AND SEQUENCING

# ESSENTIAL TECHNIQUES SERIES

*Series Editor*
**D. Rickwood**
Department of Biological and Chemical Sciences, University of Essex, Wivenhoe Park, Colchester, UK

*Published titles*
**Antibody Applications**
**Gel Electrophoresis: Nucleic Acids**
**DNA Isolation and Sequencing**

*Forthcoming titles*
**PCR**
**Gel Electrophoresis: Proteins**
**Gene Transcription**
**Human Chromosome Preparation**
**Cell Biology**

# DNA ISOLATION AND SEQUENCING

**Bruce A. Roe, Judy S. Crabtree and Akbar S. Khan**

*Department of Chemistry and Biochemistry, The University of Oklahoma, Norman, OK 73019, USA*

JOHN WILEY & SONS

Chichester • New York • Brisbane • Toronto • Singapore

Published in association with BIOS Scientific Publishers Limited

NORTHWEST MISSOURI STATE
UNIVERSITY LIBRARY
MARYVILLE, MO 64468

© 1996 John Wiley & Sons Ltd, Baffins Lane, Chichester, West Sussex PO19 1UD, UK, tel (01243) 779777. Published in association with BIOS Scientific Publishers Ltd, 9 Newtec Place, Magdalen Road, Oxford OX4 1RE, UK.

All rights reserved. No part of this book may be reproduced by any means, or transmitted, or translated into a machine language without the written permission of the publisher.

**British Library Cataloguing in Publication Data**
A catalogue record for this book is available from the British Library.

0 471 96324 0

**Library of Congress Cataloging in Publication Data**

Roe, Bruce A.
    DNA isolation and sequencing / Bruce A. Roe.
        p. cm.——(Essential techniques)
    Includes bibliographical references and index.
    ISBN 0–471–96324–0 (alk. paper)
    1. Nucleotide sequence.——Methodology.   2. DNA—Separa-
        tion
—Methodology.   I. Title.   II. Series: Essential techniques series.
QP625.N89R64   1996
574.87′3282—dc20                                                        95-48844
                                                                              CIP

Typeset by Footnote Graphics Ltd, Warminster, UK
Printed and bound in UK by Biddles Ltd, Guildford, UK

The information contained within this book was obtained by BIOS Scientific Publishers Limited from sources believed to be reliable. However, while every effort has been made to ensure its accuracy, no responsibility for loss or injury occasioned to any person acting or refraining from action as a result of the information contained herein can be accepted by the publishers, authors or editors.

SEP 2 0 1996

574.8732
P492

# CONTENTS

# ABBREVIATIONS

| | |
|---|---|
| Amp | ampicillin |
| APS | ammonium persulfate |
| BSA | bovine serum albumin |
| $c^7dGTP$ | 7-deaza-dGTP |
| CIAP | calf intestinal alkaline phosphatase |
| CPG | controlled-pore glass |
| DMF | dimethylformamide |
| DMSO | dimethylsulfoxide |
| DMT | dimethyloxytrityl |
| DTT | dithiothreitol |
| EDTA | ethylenediamine tetraacetate |
| EtBr | ethidium bromide |
| FSB | frozen storage buffer |
| IPTG | isopropyl β-D-thiogalactopyranoside |
| KDB | Klenow dilution buffer |
| MTBE | modified Tris-borate EDTA |
| PCR | polymerase chain reaction |
| PEG | polyethylene glycol |
| RF | replicative form |
| SDS | sodium dodecylsulfate |
| SSC | standard saline-citrate |
| STS | sequence-tagged sites |
| TB | Terrific Broth |
| TEMED | $N,N,N',N'$-tetramethylenediamine |
| Tet | tetracycline |
| Tris | Tris(hydroxymethyl)aminomethane |
| UV | ultraviolet |
| X-gal | 5-bromo-4-chloro β-D-galactopyranoside |

# PREFACE

This manual is a compilation of many of the everyday methods used in the average molecular biology laboratory, with emphasis on the techniques for large-scale DNA sequencing. Various forms of the manual have been in use in our laboratory for several years, and this latest version has been updated to include more detailed DNA sequencing protocols and DNA sequencing automation techniques.

This manual has been written in a protocol format, with little theoretical discussion. For theory and additional information, users of this manual are referred back to the original literature, or to other textual manuals such as that published by Sambrook, Fritsch and Maniatis (1989) see reference 1, Chapter 1.

The following persons are acknowledged for contributing methods and suggestions during the assembly of this manual: Stephanie Chissoe, Sandy Clifton, Dennis Burian, Rick Wilson, Din-Pow Ma, James Wong, Leslie Johnston-Dow, Elaine Mardis, Zhili Wang, Kala Iyer, Steve Toth, Guozhang Zhang, Hua Qin Pan and other members of the Roe laboratory, both past and present.

*Bruce A. Roe*

# SAFETY

Attention to safety aspects is an integral part of all laboratory procedures and national legislations impose legal requirements on those persons planning or carrying out such procedures. While the authors, editor and publisher believe that the recipes and practical procedures, as set forth in this book, are in accord with current recommendations and practice at the time of publication, they accept no legal responsibility for any errors or omissions, and make no warranty, expressed or implied, with respect to material contained herein. It remains the responsibility of the reader to ensure that the procedures which are followed are carried out in a safe manner and that all necessary safety instructions and national regulations are implemented.

In view of ongoing research, equipment modifications and changes in governmental regulations, the reader is urged to review and evaluate the information provided by the manufacturer, for each reagent, piece of equipment or device, for any changes in the instructions or usage and for added warnings and precautions.

These are freely accessible using:
http://joule.pcl.ox.ac.uk/MSDS/.
Other safety infomation on the Internet can be accessed on:
gopher://atlas.chem.utah.edu/11/MSDS
gopher://ginfo.cs.fit.edu:70/lm/safety/msds
http://physchem.ox.ac.uk/MSDS
http://www.fisher1.com/Fischer/Alphabetical Index.html
http://www.pp.orst.edu

You are actively encouraged to check these data sheets to confirm our assignments and for more detailed information on individual hazards; however the author, editor and publisher can accept no responsibility for any material contained in these data sheets. Furthermore, you must always follow the precautions outlined on labels and data sheets provided by individual manufacturers.

**Radiation.** The use of radioisotopes is subject to legislation and requires permission in most countries. Furthermore, national guidelines for their use and disposal must be rigorously adhered to. The procedures in protocols that use radioisotopes must only be carried out by individuals

All procedures mentioned within this book must be carried out under conditions of good laboratory practice in accordance with local and national guidelines. Some procedures involve specific hazards, including but not limited to hazards in the following categories:

**Chemical.** A number of the reagents are known to be carcinogenic, mutagenic, toxic, inflammable, highly reactive or otherwise hazardous. Substances known to be hazardous have been marked with the symbol ⚠ in the list of reagents (but not subsequently) for each protocol, or if they appear as alternatives to the main protocol, the *first time* they appear in the notes. The reader should consult the safety notes on these pages before embarking on any of the procedures covered. This is in no way meant to imply that undesignated chemicals are nonhazardous, and all laboratory chemicals should be handled with extreme caution. Information is not available on the possible hazards of many compounds. The criteria we have generally used for denoting a substance with ⚠ is based upon a hazard level of 2 or more (on a scale 0–4) in any of the categories in the Baker Saf-T-Data™ system used in the material safety data sheets (MSDS) held at the University of Oxford, UK.

who have received training in the use of such material using the appropriate facilities, protection and personal monitoring procedures.

**Biological.** Antibodies, sera and cells (particularly, but not exclusively, those of human and nonhuman primate origin) pose a significant biological hazard. All such materials, whatever their origin, may harbor human pathogens and should be handled as potentially infectious material in accordance with local guidelines. Any recombinant DNA work associated with protocols is likely to require permission from the relevant regulatory body and you must consult your local safety officer before embarking upon this work.

**Electrical**. Many of the procedures in this book use electrical equipment. Electrophoresis techniques may present particular hazards of this nature.

**Lasers.** Flow cytometers and certain other types of laboratory equipment contain lasers. Users should ensure they are fully aware of the potential hazards of using such equipment.

# I  GENERAL METHODS

## Methods available

### *Phenol extraction of DNA samples* (see *Protocol 1*)

Phenol extraction is a common procedure used to purify a DNA sample [1]. Typically, an equal volume of TE-saturated phenol is added to an aqueous DNA sample in a microcentrifuge tube. The mixture is vortexed vigorously, and then centrifuged to achieve phase separation. The upper, aqueous layer is removed carefully to a new tube, avoiding the phenol interface, and is then subjected to two ether extractions to remove residual phenol. An equal volume of water-saturated ether is added to the tube, the mixture is vortexed and the tube is centrifuged to allow phase separation. The upper, ether layer is removed and discarded, including phenol droplets at the interface. After this extraction is repeated, the DNA is concentrated by ethanol precipitation.

### *Concentration of DNA by ethanol precipitation* (see *Protocol 2*)

Typically, 2.5–3 volumes of an ethanol/acetate solution is added to the DNA sample in a microcentrifuge tube, which is placed in an ice-water bath for at least 10 min. Frequently, this precipitation is performed by incubation at −20°C overnight [1]. To recover the precipitated DNA, the tube is centrifuged, the supernatant discarded, and the DNA pellet is rinsed with a more dilute ethanol

## References

1. Sambrook, J., Fritsch, E.F. and Maniatis, T. (1989) *Molecular Cloning: A Laboratory Manual*. Cold Spring Harbor Laboratory Press, New York.
2. Studier, F.W. (1973) *J. Mol. Biol.* **79**:237.
3. Richardson, C.C. (1971) In *Procedures In Nucleic Acids Research*, Vol 2 (G.L. Cantoni and D.R. Davies, eds), p. 815. Harper and Row, New York.
4. Yanisch-Perron, C., Vieira, J. and Messing, J. (1985) *Gene*, **33**:103.
5. Jerpseth, J., Greener, A., Short, J.M., Viola, J. and Kretz, P.L. (1992) *Strategies*, **5**:81.
6. Woodcock, D.M., Crowther, D.M., Doherty, J., Jefferson, S., DeCruz, E., Noyer-Weidner, M., Smith, S.S., Michael, M.Z. and Graham, M.W. (1989) *Nucleic Acids Res.* **17**:3469.
7. Murray, N.E., Brammar, W.J. and Murray, K. (1977) *Mol. Gen. Genet.* **150**:53.
8. Messing, J. (1983) *Methods Enzymol.* **101**:20.
9. Bachman, B.J. (1990) *Microbiol. Rev.* **54**:130.

solution. After a second centrifugation, the supernatant again is discarded, and the DNA pellet is dried in a Savant Speed-Vac.

## Restriction digestion (see *Protocol 3*)

Restriction enzyme digestions are performed by incubating double-stranded DNA molecules with an appropriate amount of restriction enzyme, in its respective buffer as recommended by the supplier, and at the optimal temperature for the specific enzyme. The optimal sodium chloride concentration in the reaction varies for different enzymes, and a set of three standard buffers containing three concentrations of NaCl are prepared and used when necessary. Typical digestions include a unit of enzyme per microgram of starting DNA, and one enzyme unit usually (depending on the supplier) is defined as the amount of enzyme needed to digest one microgram of double-stranded DNA completely in 1 h at the appropriate temperature. These reactions usually are incubated for 1–3 h, to ensure complete digestion, at the optimal temperature for enzyme activity, typically 37°C.

## Agarose gel electrophoresis (see *Protocol 4*)

Agarose gel electrophoresis [2] is employed to check the progression of a restriction enzyme digestion, to determine quickly the yield and purity of a DNA isolation or polymerase chain reaction (PCR) and to size-fractionate DNA molecules, which then could be eluted from

## Protocols provided

1. *Phenol extraction of DNA samples*
2. *Concentration of DNA by ethanol precipitation*
3. *Restriction digestion*
4. *Agarose gel electrophoresis*
5. *Elution of DNA fragments from agarose*
6. *Kinase end-labeling of DNA*
7. *Bacterial cell maintenance and storage*
8. *Fragment purification on Sephacryl S-500 spin columns*

the gel. Prior to gel casting, dried agarose is dissolved in buffer by heating, and the warm gel solution then is poured into a mold (made by wrapping clear tape around and extending above the edges of an 18 cm × 18 cm glass plate), which is fitted with a well-forming comb. The percentage of agarose in the gel varies. Although 0.7% agarose gels typically are used, in cases where the accurate size-fractionation of DNA molecules smaller than 1 kb is required, a 1%, 1.5% or 2% agarose gel is prepared, depending on the expected size(s) of the fragment(s). Ethidium bromide is included in the gel matrix to enable fluorescent visualization of the DNA fragments under ultraviolet (UV) light. Agarose gels are submerged in electrophoresis buffer in a horizontal electrophoresis apparatus. The DNA samples are mixed with gel tracking dye and loaded into the sample wells. Electrophoresis is usually at 150–200 mA for 0.5–1 h at room temperature, depending on the desired separation. When low-melting agarose is used for preparative agarose gels, electrophoresis is at 100–120 mA for 0.5–1 h, again depending on the desired separation, and a fan is positioned such that the heat generated is dissipated rapidly. Size markers are electrophoresed with DNA samples when appropriate for fragment size determination. Two size markers are used, πX174 cleaved with restriction endonuclease HaeIII to identify fragments between 0.3 and 2 kb and λ phage cleaved with restriction endonuclease HindIII to identify fragments between 2 and 23 kb. After electrophoresis, the gel is placed on a UV light box

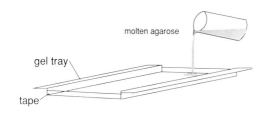

molten agarose

gel tray

tape

**Figure 1.** Preparation of a mold tray for agarose gels using plastic tape. Figure reproduced from Jones, P., Qui, J. and Rickwood, D. (1994) *RNA Isolation and Analysis* published by BIOS Scientific Publishers, Oxford.

and a picture of the fluorescent ethidium bromide-stained DNA separation pattern is taken with a Polaroid camera.

### Elution of DNA fragments from agarose (see *Protocol 5*)

DNA fragments are eluted from low-melting temperature agarose gels using an unpublished procedure first developed in our laboratory. Here, the band of interest is excised with a sterile razor blade, placed in a microcentrifuge tube, frozen at $-70°C$ and then melted. Then, TE-saturated phenol is added to the melted gel slice, and the mixture is again frozen and then thawed. After this second thawing, the tube is centrifuged and the aqueous layer removed to a new tube. Residual phenol is removed with two ether extractions, and the DNA is concentrated by ethanol precipitation.

### Kinase end-labeling of DNA (see *Protocol 6*)

Typical 5′-kinase labeling reactions include the DNA to be labeled, $[\gamma\text{-}^{32}P]dATP$, T4 polynucleotide kinase and buffer [3]. After incubation at $37°C$, reactions are heat inactivated by incubation at $80°C$. Portions of the reactions are mixed with gel loading dye and loaded into a well of a polyacrylamide gel and electrophoresed. The gel percentage and electrophoresis conditions vary depending on the sizes of the DNA molecules of interest. After electrophoresis, the gel is· dried and exposed to X-ray film for radiolabeled DNA sequencing.

***Bacterial cell maintenance and storage*** (see *Protocol 7*)

Three strains of *Escherichia coli* are used in these studies: JM101 for M13 infection and isolation [4], XL1BMRF′ (Stratagene) for M13 or pUC-based DNA transformation [5] and ED8767 for cosmid DNA transformation [6, 7 and H. Revel, personal communication]. To maintain their respective F′ episomes necessary for M13 viral infection [8], JM101 is streaked on to a M9 minimal media (modified from that given in ref. 1) plate and XL1BMRF′ is streaked on to an LB plate [1] containing tetracycline. ED8767 is streaked on to an LB plate containing ampicillin. These plates are incubated at 37°C overnight. For each strain, 3 ml of appropriate liquid media are inoculated with a smear of several colonies and incubated at 37°C with shaking at 250 r.p.m. for 8 h, and those cultures are then transferred into 50 ml of respective liquid media and incubated for an additional 12–16 h. Glycerol is added to a final concentration of 20%, and the glycerol stock cultures are distributed in 1.3 ml aliquots and frozen at −70°C until use [1].

***Fragment purification on Sephacryl S-500 spin columns***
(see *Protocol 8*)

DNA fragments larger than a few hundred base pairs can be separated from smaller fragments by chromatography on a size exclusion column such as Sephacryl S-500. To simplify this procedure, a mini-spin column method has been developed.

## Notes on precipitation of nucleic acids

### General rules
Most nucleic acids may be precipitated by addition of monovalent cations and 2–3 vol. of cold 95% ethanol, followed by incubation at 0 to −70°C. The DNA or RNA may then be pelleted by centrifugation at 10 000–13 000 r.p.m. for 15 min at 4°C. A subsequent wash with 70% ethanol, followed by brief centrifugation, removes residual salt.

The general procedure for precipitating DNA and RNA is:

1  Add 0.1 vol. of 3 M sodium acetate, pH 4.5 to the nucleic acid solution to be precipitated, 5 M ammonium acetate, pH 7.4, NaCl and LiCl may be used as alternatives to sodium acetate. DNA also may be precipitated by addition of 0.6 vol. of iso-propanol.
2  Add 2 vol. of cold 95% ethanol⚠.
3  Place at −70°C for at least 30 min, or at −20°C overnight.

or alternatively:

1  Combine 95 ml of 100% ethanol⚠ with 4 ml of 3 M sodium acetate, pH 4.5 and 1 ml of sterile water. Mix by inversion and store at −20°C.
2  Add 2.5 vol. of cold ethanol/acetate solution⚠ to the nucleic acid solution to be precipitated.
3  Place at −70°C for at least 30 min or −20°C for 2 h to overnight.

### Oligonucleotides
Add 0.1 vol. of 3 M sodium acetate, pH 4.5, and 3 vol. of cold 95% ethanol⚠. Place at −70°C for at least 1 h.

## RNA

Add 0.1 vol. of 1 M sodium acetate, pH 4.5, and 2.5 vol. of cold 95% ethanol. Precipitate large volumes at −20°C overnight. Small volume samples may be precipitated by placing in powdered dry ice or a dry ice–ethanol bath for 5–10 min.

## Isobutanol concentration of DNA

DNA samples may be concentrated by extraction with isobutanol. Add slightly more than one volume of isobutanol, vortex vigorously and centrifuge to separate the phases. Discard the isobutanol (upper) phase, and extract once with water-saturated diethyl ether to remove residual isobutanol. The nucleic acid then may be ethanol precipitated as described previously.

## Notes on phenol extraction of nucleic acids

The standard and preferred way to remove proteins from nucleic acid solutions is by extraction with neutralized phenol or phenol:chloroform. Generally, samples are extracted by addition of 1.0 vol. of neutralized (with TE buffer, pH 7.5) phenol to the sample, followed by vigorous mixing for a few seconds to form an emulsion. Following centrifugation for a few minutes, the aqueous (top) phase containing the nucleic acid is recovered and transferred to a clean tube. Residual phenol is then removed by extraction with an equal volume of water-saturated diethyl ether. Following centrifugation to separate the phases, the ether (upper) phase is discarded and the nucleic acid is ethanol precipitated as described previously.

A 1:1 mixture of phenol and chloroform is also useful for the removal of protein from nucleic acid samples. Following extraction with phenol:chloroform, the sample should be extracted once with an equal volume of chloroform, and ethanol precipitated as described above.

*General methods*

# Protocol 1. **Phenol extraction of DNA samples**

## Reagents

95% Ethanol/0.12 M sodium acetate (ethanol/acetate⚠)
TE-saturated phenol⚠: add an equal amount of 10 mM Tris–HCl,
  pH 7.5–8.0, 1 mM Na$_2$EDTA, pH 8.0 to ultrapure phenol①,
  mix well, allow phases to separate, remove and discard upper
  (aqueous) phase. Repeat until the pH of the aqueous phase is
  between 7.5 and 8.0 (store at 4°C)
1.0 M Tris–HCl, pH 7.6

1.0 M Tris–HCl, pH 8.0
Water-saturated diethyl ether

## Equipment

Microcentrifuge
1.5 ml Microcentrifuge tubes
Vortex mixer

## Procedure

1  Add an equal volume of TE-saturated phenol to the DNA sample
   contained in a 1.5 ml microcentrifuge tube and vortex for 15–30 sec.

2  Centrifuge the emulsion for 5 min at 12 000 r.p.m. at room temperature
   to separate the phases.

3  Remove about 90% of the upper, aqueous layer to a clean tube,
   carefully avoiding proteins. Add an equal volume of 1:1 TE-saturated
   phenol:chloroform, centrifuge and remove to a clean tube as above.
   This additional extraction is not usually necessary if care is taken during
   the first phenol extraction.

4  Add an equal volume of water-saturated ether to the phenol extracted

## Notes

This procedure will take about 15 min.

①  Pure phenol is colorless, a pinky color indicates oxidation
   and in this case the phenol needs to be distilled before use.

aqeous layer, vortex briefly and centrifuge for 3 min at room temperature. Remove and discard the upper, ether layer, taking care to remove phenol droplets at the ether:aqueous interface. Repeat the ether extraction.

5  Ethanol-precipitate the DNA by adding 2.5–3 vol. of ethanol/acetate, as discussed in *Protocol 2*.

*Protocol 1.  Phenol extraction of DNA samples*

# Protocol 2. Concentration of DNA by ethanol precipitation

## Reagents

0.5 M EDTA, pH 8.0
70% Ethanol⚠
95% Ethanol/0.12 M sodium acetate (ethanol/acetate)
10 M NaOH⚠
TE (10:0.1) buffer
1.0 M Tris–HCl, pH 7.6

## Equipment

Ice-water bath
Microcentrifuge
0.5 ml Microcentrifuge tubes
1.5 ml Microcentrifuge tubes
Paper towels
Savant Speed-Vac

## Procedure

1 Add 2.5–3 vol. of ethanol/acetate to the DNA sample contained in a 1.5 ml microcentrifuge tube, invert to mix, and incubate in an ice-water bath for at least 10 min. [1]

2 Centrifuge at 12 000 r.p.m. in a microcentrifuge for 15 min at 4°C, decant the supernatant, and drain the tube inverted on a paper towel.①

3 Add 70% ethanol (corresponding to ~2 vol. of the original sample), incubate at room temperature for 5–10 min, centrifuge again for 5 min and then decant and drain the tube, as above.

4 Place the tube in a Savant Speed-Vac and dry the DNA pellet for about 5–10 min or until dry.

## Notes

This procedure takes about 1 hour.

① If precipitating submicrogram amounts of DNA, a better yield is achieved by ultracentrifugation.

5 Dissolve the dried DNA in TE (10:0.1) buffer.

6 It is advisable to aliquot the DNA purified in large-scale isolations (i.e. 100 µg or more) into several small (0.5 ml) microcentrifuge tubes for frozen storage to decrease excessive freezing and thawing cycles.

## Pause points

[1] The sample may be left at $-20°C$ overnight at this stage.

*Protocol 2. DNA concentration by ethanol precipitation*

# Protocol 3. **Restriction digestion**

## Reagents

10× Assay buffer: refer to the supplier's catalog for the chart of enzyme activity in a range of salt concentrations to choose the appropriate assay buffer (10× high, 10× medium, or 10× low salt buffer, or 10× *Sma*I buffer for *Sma*I digestions)

φ174/*Hae*III marker (Pharmacia)

Restriction enzyme (purchased from Bethesda Research Laboratories, New England Biolabs, or United States Biochemicals)

## Equipment

Agarose gel electrophoresis apparatus
Automatic pipettor
Microcentrifuge tube

## Procedure

1 Prepare the reaction for restriction digestion by adding the following reagents, in the order listed, to a 0.5 ml microcentrifuge tube:
   - z/10 of 10x assay buffer
   - x μl of DNA
   - y μl (1–10 units per μg DNA) of restriction enzyme(1)
   - z μl of sterile double-distilled water typically to a final volume of 20 μl.
   The volume of the reaction components (x, y and z) depends on the amount and size of the DNA to be digested. Larger DNAs should be digested in larger total volumes (between 50 and 100 μl), as should greater amounts of DNA.

2 Gently mix by pipetting and incubate the reaction at the appropriate temperature (typically 37°C) for 1–3 h.

## Notes

This procedure takes about 3.5 hours.

(1) If desired, more than one enzyme can be included in the digest if both enzymes are active in the same buffer and at the same incubation temperature.

3 Inactivate the enzyme(s) by heating at 70–100°C for 10 min or by phenol extraction (see the supplier's catalog to determine the degree of heat inactivation for a given enzyme). Prior to use in further protocols, such as dephosphorylation or ligation, an aliquot of the digestion should be assayed by agarose gel electrophoresis versus nondigested DNA and a size marker, if necessary.

## Protocol 4. Agarose gel electrophoresis

### Reagents

Agarose: genetic technology grade (800669) or low melting
temperature (800259) agarose from Schwann/Mann Biotech
10× Agarose gel loading dye
5 mg/ml Ethidium bromide (EtBr)⚠
20× TAE buffer

### Equipment

Agarose gel electrophoresis apparatus
500 ml Erlenmeyer flask
Fan
Long-wave UV light box⚠
Microwave
Polaroid camera
Taped plate with casting combs

### Procedure

1 Prepare an agarose gel, according to the recipes listed below, by combining the agarose (low-melting temperature agarose may also be used) and water in a 500 ml Erlenmeyer flask, and heating in a microwave for 2–4 min until the agarose is dissolved.

|                        | 0.7%     | 1.0%     | 2.0%     |
|------------------------|----------|----------|----------|
| Agarose                | 1.05 g   | 1.5 g    | 3.0 g    |
| 20× TAE                | 7.5 ml   | 7.5 ml   | 7.5 ml   |
| Double-distilled water | 142.5 ml | 142.5 ml | 142.5 ml |
| EtBr (5 mg/ml)         | 25 µl    | 25 µl    | 25 µl    |
| Total volume           | 150 ml   | 150 ml   | 150 ml   |

### Notes

This procedure takes about 2.5 hours.

2 Add 20× TAE and EtBr, swirl to mix, and pour the gel on to a taped plate with casting combs in place. Allow 20–30 min for solidification.

3 Carefully remove the tape and the gel casting combs and place the gel in a horizontal electrophoresis apparatus. Add 1× TAE electrophoresis buffer (diluted from 20× stock) to the reservoirs until the buffer just covers the agarose gel.

4 Add at least 0.1 vol. of 10× agarose gel loading dye to each DNA sample, mix and load into the wells. Electrophorese the gel at 150–200 mA until the required separation has been achieved, usually 0.5–1 h (100–120 mA for low-melting temperature agarose), and cool the gel during electrophoresis with a fan. Visualize the DNA fragments on a long-wave UV light box and photograph with a Polaroid camera.

*Protocol 4. Agarose gel electrophoresis*

## Protocol 5. Elution of DNA fragments from agarose

### Reagents

0.5 M EDTA, pH 8.0
95% Ethanol
10 M NaOH△
TE-saturated phenol△
1.0 M Tris–HCl, pH 7.6
1.0 M Tris–HCl, pH 8.0
Water-saturated ether

### Equipment

Microcentrifuge
1.5 ml Microcentrifuge tubes
Vortex mixer
Water bath or dry oven

### Procedure

1 Place the excised DNA-containing agarose gel slice in a 1.5 ml microcentrifuge tube and freeze at −70°C for at least 15 min, or until frozen. [1]

2 Melt the slice by incubating the tube at 65°C.

3 Add 1 vol. of TE-saturated phenol, vortex for 30 sec and freeze the sample at −70°C for 15 min.

4 Thaw the sample, and centrifuge in a microcentrifuge at 12 000 r.p.m. for 5 min at room temperature to separate the phases. The aqueous phase is then removed to a clean tube, extracted twice with an equal volume of water-saturated ether, ethanol precipitated and the DNA pellet is rinsed and dried.

### Notes

This procedure takes about 2 hours.

### Pause points

[1] The gel slice may be left frozen at −70°C.

# Protocol 6.  **Kinase end-labeling of DNA**

## Reagents

[γ-$^{32}$P]dATP: (220 M Bq/mmol 6000 Ci/mmol)⚠①
1.0 M Dithiothreitol (DTT)
10× Kinase buffer
1.0 M MgCl$_2$

1.0 M Tris–HCl, pH 7.6
1.0 M Tris–HCl, pH  8.0
T4 polynucleotide kinase:  (U. S. Biochemicals)

## Equipment

0.5 ml Microcentrifuge tubes

## Procedure

1 Add the following reagents to a 0.5 ml microcentrifuge tube, in the order listed:
   • 1 µl of 10× kinase buffer
   • x µl of DNA (1–2 µg)②
   • 10 µCi of [γ-$^{32}$P]dATP
   • 1 µl (3 U/µl) of T4 polynucleotide kinase
   and add sterile double-distilled water to a final volume of 10 µl.

2 Incubate at 37°C for 30–60 min.

3 Heat the reaction at 65°C for 10 min to inactivate the kinase.

4 Remove unincorporated $^{32}$P-dATP by passage over a Sephacryl 500 column (see *Protocol 8*).

## Notes

This procedure takes about 1.5 hours.

① Use 8 mm Plexiglas (Perspex) shielding to reduce exposure to radiation.

② The volume of DNA should be adjusted such that 1–2 µg is present in the final reaction mixture.

**17**

# Protocol 7. Bacterial cell maintenance and storage

## Reagents

20% Glucose (filter to sterilize)
LB agar
LB medium
M-9 agar
M-9 medium (liquid)
10× M-9 salts
Sterile glycerol
Tetracycline stock (Tet) solution (10 mg/ml)
1% Thiamine

## Equipment

Autoclave
Erlenmeyer flasks
12× 75 mm Falcon tubes
100 ml Graduated cylinder capped with aluminum foil
Sterile Petri dishes
Sterile pipettes or graduated cylinders

## Procedure

1 Streak a culture of the *Escherichia coli* cell strain on to an agar plate of the respective medium listed below, and incubate at 37°C overnight.

   • XL1BMRF' (Stratagene) → LB-Tet
   • JM101　　　　　　　 → M-9
   • ED8767　　　　　　　 → LB

2 Pick several colonies into a 12× 75 mm Falcon tube containing a 2 ml aliquot of the respective liquid media, and incubate for 8–10 h at 37°C with shaking at 250 r.p.m.

## Notes

This procedure takes about 24 hours.

3  Transfer the 2-ml culture into an Erlenmeyer flask containing 50 ml of the respective liquid media and further incubate overnight (12–16 h) at 37°C with shaking at 250 r.p.m.

4  Add 12.5 ml of sterile glycerol for a final concentration of 20%, and distribute the culture in 1.3 ml aliquots into 12× 75 mm Falcon tubes.

5  Store glycerol cell stocks frozen at −70°C until use.

## Reagents

0.7% Agarose gel
0.5 M EDTA, pH 8.0
95% Ethanol
1.0 M $MgCl_2$
10 M NaOH⚠
Nebulized cosmid, plasmid or P1 DNA
φX174/*Hae*III marker (Pharmacia)
Sephacryl S-500 (Pharmacia)
TE (10:1) buffer

10× TM buffer
1 M Tris–HCl, pH 8.0

## Equipment

Agarose gel electrophoresis apparatus
1.5 ml Microcentrifuge tubes
Mini-spin column (Millipore)
Screw cap bottles or centrifuge tubes
Table-top low speed centrifuge
Vortex mixer

## Procedure

1  Thoroughly mix a fresh, new bottle of Sephacryl S-500, distribute in 10 ml portions, and store in screw cap bottles or centrifuge tubes in the cold room at 4°C.

2  Prior to use, briefly vortex the matrix and, without allowing it to settle, add 500 μl of this slurry to a mini-spin column (Millipore) which has been inserted into a 1.5 ml microcentrifuge tube.

3  Following centrifugation at 800 *g* for 2 min in a table-top centrifuge, carefully add 200 μl of 100 mM Tris–HCl (pH 8.0) to the top of the

## Notes

This procedure takes about 1.5 hours plus the time for ethanol precipitation.

Sephacryl matrix and centrifuge for 2 min at 800 *g*. Repeat this step twice more. Place the Sephacryl matrix-containing spin column into a new microcentrifuge tube.

4  Carefully add 40 µl of nebulized cosmid, plasmid or P1 DNA which has been end-repaired to the Sephacryl matrix (save 2 µl for later agarose gel analysis) and centrifuge at 1200 *g* for 5 min. Remove the column, save the solution containing the eluted, large DNA fragments (fraction 1). Apply 40 µl of 1× TM buffer (diluted from 10× stock) and recentrifuge for 2 min at 800 *g* to obtain fraction 2, and repeat this 1× TM rinse step twice more to obtain fractions 3 and 4.

5  To check the DNA fragment sizes, load 3–5 µl of each eluent fraction on to a 0.7% agarose gel that includes as controls, 1–2 µl of a φX174/*Hae*III digest and 2 µl of unfractionated, nebulized DNA saved from Step 4 above.

6  The fractions containing the nebulized DNA in the desired size ranges (typically fractions 1 and 2) are phenol extracted separately and concentrated by ethanol precipitation prior to the kinase reaction (see *Protocol 6*).

*Protocol 8.  Purification on Sephacryl S-500 spin columns*

# II  RANDOM SUBCLONE GENERATION

## Methods available

*Sonication* (see *Protocol 9*)

The generation of DNA fragments by sonication is performed by placing a microcentrifuge tube containing the buffered DNA sample into an ice-water bath in a cup-horn sonicator and sonicating for a varying number of 10-sec bursts using maximum output and continuous power [1], essentially as described by Bankier *et al.* [2]. During sonication, temperature increases result in uneven fragment distribution patterns and, for that reason, the temperature of the bath is monitored carefully during sonication, and fresh ice-water is added when necessary. The exact conditions for sonication are determined for a given DNA sample before a preparative sonication is performed. Approximately 100 μg of DNA sample, in 350 μl of buffer, is distributed into 10 aliquots of 35 μl, five of which are subjected to sonication for increasing numbers of 10-sec bursts. Aliquots from each time point are electrophoresed on an agarose gel versus the φX174 size marker [3] to determine the approximate DNA fragment size range for each sonication time point. Once optimal sonication conditions are determined, the remaining five DNA aliquots (~50 μg) are sonicated according to those predetermined conditions. After sonication, the five tubes are placed in an ice-water bath until frag-

## References

1. Bodenteich, A., Chissoe, S., Wang, Y.F. and Roe, B.A. (1994) in *Automated DNA Sequencing and Analysis Techniques* (C. Venter, ed.), p. 42. Academic Press, London.
2. Bankier, A.T., Weston, K.M. and Barrell, B.G. (1987) *Methods Enzymol.* **155**:51.
3. Bethesda Research Laboratories (1994) catalog.
4. Mandel, M. and Higa, A. (1970) *J. Mol. Biol.* **53**:154.
5. Hanahan, D. (1983) *J. Mol. Biol.* **166**:557.
6. Cohen, S.N., Chang, A.C.Y. and Hsu, L. (1972) *Proc. Natl Acad. Sci. USA*, **69**:2110.

## Protocols provided

9. *Sonication*
10. *Nebulization*
11. *Random fragment end-repair, size selection and phosphorylation*
12. *DNA ligation*
13. *Competent cell preparation*
14. *Bacterial cell transformation*

ment end-repair and size selection, discussed below.

***Nebulization*** (see *Protocol 10*)
A nebulizer can be purchased from a local supplier whose name can be obtained by calling the manufacturer, IPI Medical Products Inc. listed in Appendix C.

We follow a protocol sent to us by Steve Surzycki at the Department of Biology, Indiana University with two modifications as follows:

(i) Cover the hole where normally the mouth piece is attached with a cap QS-T from ISOLAB Inc. (Drawer 4350 Akron, OH 44303).
(ii) Leakage may occur where the hose for the nitrogen is attached. Nalgene tubing (VI grade 3/16″ i.d.) seals better than the tubing which comes with the nebulizer.

A nebulizer containing 2 ml of a buffered DNA solution (~50 µg) containing 25–50% glycerol is placed in an ice-water bath and is subjected to nitrogen gas at a pressure of 10 p.s.i. (0.7 bar) for cosmid or 30 p.s.i. (2 bar) for plasmid for 2.5 min (ref. 7 and S. J. Surzycki, personal communication). Nitrogen gas pressure is the primary determinant of DNA fragment size and, although pressure studies should be performed with each cosmid or plasmid, a pressure of 10 p.s.i. (0.7 bar) almost always results in the desired (600–1500 bp) fragment size range. As discussed above for sonication, the use of an

*Random subclone generation*

ice-water bath for nebulization also is critical for the generation of evenly distributed DNA fragments. During the nebulization process, unavoidable leaks are minimized by securely tightening the lid of the nebulizer chamber and sealing the larger hole in the top piece with a plastic cap. To prepare for fragment end-repair, the nebulized DNA typically is divided into four tubes and concentrated by ethanol precipitation (see *Protocol 2*).

### *Random fragment end-repair, size selection and phosphorylation*
(see *Protocol 11*)
Since both sonicated and nebulized DNA fragments usually contain single-stranded ends, the samples are end-repaired prior to ligation into blunt-ended vectors [1, 2]. A combination of T4 DNA polymerase and Klenow DNA polymerase are used to 'fill-in' the DNA fragments by catalyzing the 3′–5′ incorporation of complementary nucleotides into double-stranded fragments with a 5′ overhang. Additionally, the single-stranded 3′–5′ exonuclease activity of T4 DNA polymerase is used to degrade 3′ overhangs. The reactions include the two enzymes, buffer and deoxynucleotides, and are incubated at 37°C.

Following fragment end-repair, the DNA samples are electrophoresed on a preparative low-melting temperature agarose gel versus the φX174 marker and, after appropriate separation, the fragments in the size range from 600 to 1500 bp are eluted from the gel, as discussed above (see *Protocol 5*). Alternatively, the fragments can be purified by fractionation on a Sephacryl S-500 spin column also as discussed above (see *Protocol 8*). In both instances, the purified fragments are concentrated by ethanol precipitation followed by resuspension in kinase buffer, and phosphorylation using T4 polynucleotide kinase and rATP. The polynucleotide kinase is removed by phenol extraction, and the DNA fragments are concentrated by ethanol precipitation, dried, resuspended in buffer and ligated into blunt-ended cloning vectors. It should be noted that because a significant portion of nebulized DNA fragments are easily cloned without end-repair or kinase treatment, these two steps can be combined without significantly affecting the overall number of resulting transformed clones.

***DNA ligation*** (see *Protocol 12*)
DNA ligations are performed by incubating DNA fragments with appropriately linearized cloning vector in the presence of buffer, rATP and T4 DNA ligase [1, 2]. For random shotgun cloning, sonicated or nebulized fragments are ligated to either *Sma*I-linearized, dephosphorylated double-stranded M13 replicative form or pUC vector by incubation at 4°C overnight. A practical range of concentrations is determined based on the amount of initial DNA, and several different ligations, each with an amount of insert DNA

within that range, are used to determine the appropriate insert to vector ratio for the ligation reaction. In addition, several control ligations are performed to test the efficiency of the blunt-ending process, the ligation reaction and the quality of the vector [1, 2]. These usually include parallel ligations in the absence of insert DNA to determine the background clones arising from self-ligation of inefficiently dephosphorylated vector. Parallel ligations also are performed with a known blunt-ended insert or insert library, typically an *Alu*I digest of a cosmid, to ensure that the blunt-ended ligation reaction would yield sufficient insert-containing clones, independent of the repair process.

***Competent cell preparation*** (see *Protocol 13*)
For preparation of competent bacterial cells [4], a glycerol cell culture stock of the respective *E. coli* strain is thawed and added to 50 ml of liquid media. This culture then is pre-incubated at 37°C for 1 h, transferred to an incubator–shaker, and is incubated further for 2–3 h. The cells are pelleted by centrifugation, resuspended in $CaCl_2$ solution and incubated in an ice-water bath. After another centrifugation step, the resulting cell pellet is again resuspended in $CaCl_2$ solution to yield the final competent cell suspension. Competent cells can be stored at 4°C for up to 1 week.

Alternatively, it is possible to store the competent cells for at least 1–2 years without losing any transformation efficiency by freezing them [5]. This saves at least 4 h, since the cells will not have to be grown every time competent cells are needed.

### *Bacterial cell transformation* (see *Protocol 14*)

For DNA transformation [4, 6], the entire DNA ligation reaction is added to an aliquot of competent cells, mixed gently and incubated in an ice-water bath. This mixture is then heat-shocked briefly in a 42°C water bath for 2–5 min. At this point in the transformation, the method varies slightly depending on whether the cloning vector is M13-based or pUC-based.

For M13-based transformation [4], an aliquot of noncompetent cells is added to the heat-shocked mixture, along with the *lac* operon inducer homolog, isopropyl $\beta$-D-thiogalactopyranoside (IPTG: Sigma), and the $\beta$-galactosidase chromogenic substrate, X-gal (5-bromo-4-chloro-$\beta$-D-galactopyranoside: Sigma). Melted top agar is added, and the transformation mixture is then poured on to the surface of an agar plate. After the top agar solidifies, the plates are inverted and incubated overnight at 37°C.

For pUC-based transformation [6], an aliquot of liquid media is added to the heat-shocked mixture, which then is incubated in a 37°C water bath for 15–20 min. After recovery, the cell suspension is concentrated by centrifugation and then gently resuspended in a smaller volume of fresh liquid media. IPTG and X-gal are added to the cell mixture, which is spread on to the surface of an ampicillin-containing agar plate. After the cell mixture has diffused into the agar medium, the plates are inverted and incubated overnight at 37°C.

# Protocol 9. **Sonication**

## Reagents

1.0 M MgCl$_2$
10× TM buffer
1.0 M Tris–HCl, pH 8.0
φX174/*Hae*III marker (Pharmacia 15611-015)

## Equipment

Agarose gel electrophoresis apparatus
Heat Systems Ultrasonics W-375 cup-horn sonicator
Ice-water bath
1.5 ml Microcentrifuge tubes
Table-top microfuge

## Procedure

1 Prepare the following DNA dilution, and aliquot 35 μl into ten 1.5 ml microcentrifuge tubes:

- 100 μg of DNA
- 35 μl of 10× TM buffer
- Sterile double-distilled water to a final volume of 350 μl.

2 To determine the optimal sonication conditions, sonicate the DNA samples in five of the tubes in a Heat Systems Ultrasonics W-375 cup-horn sonicator set on 'HOLD', 'CONTINUOUS' and maximum 'OUTPUT CONTROL' = 10 under the following conditions:

## Notes

This procedure takes about 1 hour.

| Tube | No. of 10-sec bursts |
|------|----------------------|
| 1    | 1                    |
| 2    | 2                    |
| 3    | 3                    |
| 4    | 4                    |
| 5    | 5                    |

3 Cool the DNA samples by placing the tubes in an ice-water bath for at least 1 min between each 10-sec burst. Replace the ice-water bath in the cup-horn sonicator between each sample.

4 Centrifuge to collect the sample and electrophorese a 10 µl aliquot from each sonicated DNA sample on an agarose gel versus the φX174/*Hae*III size marker.

5 Based on the fragment size ranges detected from agarose gel electrophoresis, sonicate the remaining five tubes according to the optimal conditions and then place the tubes in an ice-water bath.

*Protocol 9. Sonication*

# Protocol 10. **Nebulization**

## Reagents

1.0 M MgCl$_2$
Sterile glycerol
10× TM buffer
1.0 M Tris–HCl, pH 8.0

## Equipment

Ice-water bath
1/4 inch internal diameter length of Tygon tubing
Low-speed table-top centrifuge
1.5 ml Microcentrifuge tubes
Nebulizer (IPI Medical Products)
Plastic stopper (Isolab)
Source of compressed nitrogen⚠
Styrofoam

## Procedure

1 Modify a nebulizer by removing the plastic cylinder drip ring, cutting off the outer rim of the cylinder, inverting it and placing it back into the nebulizer. Seal the large hole in the top cover (where the mouth piece was attached) with a plastic stopper and connect Tygon tubing (which eventually should be connected to a compressed nitrogen source) to the smaller hole.

2 Prepare the following DNA sample and place in the nebulizer cup:
- 50 µg of DNA
- 200 µl of 10× TM buffer
- 0.5–1 ml of sterile glycerol
- Sterile double-distilled water to a final volume of 2.0 ml.

## Notes

This procedure takes about 1 hour.

3 Nebulize in an ice-water bath at 30 p.s.i. (2 bar) for 2.5 min for plasmid, or 10 p.s.i. (0.7 bar) for 2.5 min for cosmid.

4 Briefly centrifuge at 2500 r.p.m. to collect the sample by placing the entire unit in the rotor bucket of a table-top centrifuge fitted with pieces of Styrofoam to cushion the plastic nebulizer.

5 Distribute the sample into four 1.5 ml microcentrifuge tubes and ethanol precipitate (see *Protocol 2*). Resuspend the dried DNA pellets in 35 μl of 1× TM buffer (from 10× stock) prior to proceeding with fragment end-repair (see *Protocol 11*).

# Protocol 11. Random fragment end-repair, size selection and phosphorylation

## Reagents

10× Agarose gel loading dye
10× Denaturing buffer
1.0 M Dithiothreitol (DTT)
20 mM dNTP stocks: 80 µl of 100 mM dNTP, 40 µl of TE (50:1) buffer, 280 µl of double-distilled water
0.5 M EDTA, pH 8.0
95% Ethanol
10× Kinase buffer
Klenow DNA polymerase (New England Biolabs)
1.0 M $MgCl_2$
10 M NaOH⚠

T4 Polynucleotide kinase (United States Biochemicals)
100 mM rATP (aliquot and store at −20°C)
100 mM Spermidine
TE (10:0.1) buffer
1.0 M Tris–HCl, pH 7.6
1.0 M Tris–HCl, pH 9.5

## Equipment

Agarose gel electrophoresis apparatus
Ice-water bath
1.0% Low gel temperature agarose gel (see *Protocol 4*)

## Procedure

1 To each tube containing 35 µl of DNA fragments (five of sonicated DNA and four of nebulized DNA), add:
   - 2 µl of 0.25 mM dNTPs
   - 3 µl (3 U/µl) of T4 DNA polymerase
   - 2 µl (5 U/µl) of Klenow DNA polymerase
   to give a total of 42 µl.

2 Incubate at room temperature for 30 min.

## Notes

This procedure takes about 3 hours.

3 Add 5 µl of agarose gel loading dye and apply to separate wells of a 1% low-melting temperature agarose gel and electrophorese for 30–60 min at 100–120 mA.

4 Elute the DNA from each sample lane, ethanol precipitate and resuspend the dried DNA in 36 µl of sterile, double-distilled water and add 4 µl of 10× denaturing buffer. There should be five tubes for sonicated fragments and four tubes for nebulized fragments.

5 Incubate at 70°C for 10 min, and place the samples in an ice-water bath.

6 Add the following reagents for the kinase reaction and incubate at 37°C for 10–30 min:
- 1 µl of 10 mM rATP (from 100 mM stock)
- 5 µl of 10× kinase buffer
- 1 µl (30 U/µl) of T4 polynucleotide kinase
to give a total volume of 47 µl.

7 Pool the kinase reactions, phenol extract, ethanol precipitate and resuspend the dried DNA fragments in 10 µl of TE (10:0.1) buffer. This yields a typical concentration of 500–1000 ng/µl.

*Alternatively:*

1 Resuspend DNA in 25 µl of 1× TM buffer. Add the following:
- 5 µl of 10× kinase buffer
- 5 µl of 10 mM rATP
- 7 µl of 0.25 mM dNTPs
- 1 µl (30 U/µl) of T4 polynucleotide kinase

*Protocol 11.  Random fragment end-repair, size selection and phosphorylation*

- 2 μl (5 U/μl) of Klenow DNA polymerase
- 2 μl (3 U/μl) of T4 DNA polymerase

to give a total volume of 47 μl.

2  Incubate at 37°C for 30 min.

3  Add 5 μl of agarose gel loading dye and apply to separate wells of a 1% low-melting temperature agarose gel and electrophorese for 30–60 min at 100–120 mA (see *Protocol 4*).

4  Elute the DNA from each sample lane, ethanol precipitate and resuspend in 10 μl of TE (10:0.1) buffer.

# Protocol 12. **DNA ligation**

## Reagents

1 mg/ml Bovine serum albumin (BSA) (aliquot and store at −20°C)

Cloning vector: typically *Sma*I-linearized, calf intestinal alkaline phosphatase (CIAP)-dephosphorylated pUC vector (Pharmacia), and prepared as described below. In some instances, *Sma*I-linearized, CIAP-dephosphorylated M13RF is used. 5% Polyethylene glycol (PEG) can be included in the reaction to improve the ligation efficiency slightly

1.0 M Dithiothreitol (DTT) (aliquot and store at −20°C)

Known blunt-ended insert

10× Ligation buffer

1.0 M $MgCl_2$

5% Polyethylene glycol (PEG) 8000

100 mM rATP (aliquot and store at −20°C)

T4 DNA ligase (NEB 202L)

1.0 M Tris−HCl, pH 7.6

## Equipment

1.5 ml Microcentrifuge tubes

## Procedure

1  Combine the following reagents in a microcentrifuge tube, and incubate overnight at 4°C:
    - 100–1000 ng of DNA fragments
    - 2 µl (10 ng/µl) of cloning vector
    - 1 µl of 10× ligation buffer
    - 1 µl (400 U/µl) of T4 DNA ligase
    - Sterile double-distilled water to a final volume of 10 µl.

2  Include control ligation reactions with no insert DNA and with a known blunt-ended insert (such as *Alu*I-digested cosmid).

## Notes

This procedure takes about 12 hours.

# Protocol 13. **Competent cell preparation**

## Reagents

50 mM CaCl$_2$
100% Dimethylsulfoxide (DMSO)⚠
Frozen glycerol stock of appropriate strain of *E. coli*
Frozen storage buffer (FSB)
2× TY medium

## Equipment

Autoclave
Erlenmeyer flask
Falcon culture tubes (12× 75 mm)
Ice-water bath
Sterile 50 ml polypropylene centrifuge tubes
Table-top centrifuge or high speed centrifuge with a fixed-angle rotor
37°C Water bath

## Procedure

1  Thaw a frozen glycerol stock of the appropriate strain of *E. coli*, add it to an Erlenmeyer flask containing 50 ml of pre-warmed 2× TY medium, and pre-incubate in a 37°C water bath for 1 h with no shaking. Further incubate for 2–3 h at 37°C with shaking at 250 r.p.m.

2  Transfer 40 ml of the cells to a sterile 50 ml polypropylene centrifuge tube, and collect the cells by centrifugation at 1500 *g* for 8 min at 4°C in a low speed centrifuge or 5000 *g* for 8 min at 4°C in a high speed centrifuge (DuPont) equipped with a fixed-angle rotor. For M13-based transformation, save the remaining 10 ml of culture in an ice-water bath for later use.

## Notes

This procedure takes about 5 hours.

3 After centrifugation, decant the supernatant and resuspend the cell pellet in 0.5 vol. (20 ml) of cold, sterile 50 mM $CaCl_2$, incubate in an ice-water bath for 20 min and centrifuge as before.

4 Decant the supernatant and gently resuspend the cell pellet in 0.1 vol. (4 ml) of cold, sterile 50 mM $CaCl_2$ to yield the final competent cell suspension.

### Preparation of competent cells for frozen storage

1 Inoculate 50 ml of $2 \times$ TY with a 2 ml overnight culture of GM272 and incubate in the shaker at $37°C$ for 2–2.5 h.

2 Chill the cells on ice for 10–15 min, transfer to a 50 ml polypropylene centrifuge tube and spin at 4000 $g$ (6000 r.p.m.) for 5 min at $4°C$.

3 Resuspend the cell pellet in 20 ml of FSB by gentle vortexing and incubate on ice for 10–15 min.

4 Centrifuge as before, and resuspend the cell pellet in 4 ml of FSB. Place on ice.

5 Add 140 µl of fresh DMSO and incubate on ice for 15 min. Add an additional 140 µl of DMSO and keep on ice for a further 5 min.

6 Transfer the cells to sterile Falcon culture tubes (210 µl per tube). [1]

7 To use competent cells for transformation, remove them from the freezer and thaw at $37°C$. Place on ice, add plasmid DNA and incubate for 1 h as in the standard transformation procedure given in *Protocol 14*.

## Pause points

[1] The competent cells may be placed at $-70°C$ and stored indefinitely.

*Protocol 13. Competent cell preparation*

# Protocol 14. Bacterial cell transformation

## Reagents

0.5% Ampicillin (Amp): add to media for final concentration of
   100 µg/ml
2% 5-Bromo-4-chloro-3-indolyl β-D-galactopyranoside (X-gal) in
   dimethylformamide (DMF) (aliquot and store protected from
   light at −20°C)
Competent cells
2.5% Isopropyl β-D-thiogalactopyranoside (IPTG) (aliquot and
   store at −20°C)
Lambda top agar
Noncompetent cells
1% Tetracycline stock (Tet) in 50% ethanol
2× TY medium

## Equipment

Autoclave
12× 75 Falcon tubes
Ice-water bath
Lambda plates containing antibiotic
Low speed table-top centrifuge
Pre-warmed lambda agar plate
Pre-warmed LB-Amp plate
Sterile bent glass rod or inoculating loop
Sterile glass spreader
Vortex mixer

## Procedure

1 Add the entire ligation reaction to a 12× 75 Falcon tube containing
   0.2–0.3 ml of competent cells, mix gently and incubate in an ice-water
   bath for 40–60 min. (For re-transformation of recombinant DNA, add
   ~10–100 ng of DNA directly to competent cells.)

2 Heat shock the cells by incubation at 42°C for 2–5 min.

## Notes

This procedure takes about 2 hours.

### For M13-based transformation

3 Add the following reagents to the heat-shocked transformation mixture:
- 0.2 ml of noncompetent cells
- 25 µl of IPTG (25 mg/ml in water)
- 25 µl of X-gal (20 mg/ml in DMF)
- 2.5 ml of lambda top agar.

4 Mix by vortexing briefly, and then quickly pour on to the surface of a pre-warmed lambda agar plate.

5 Allow 10–20 min for the agar to harden, and then invert and incubate overnight at 37°C.

### For pUC-based transformation

3 To the heat-shocked transformation mixture, add 1 ml of fresh 2× TY and incubate in a 37°C water bath for 15–30 min.

4 Collect the cells by centrifugation at 15 000 $g$ (3000 r.p.m.) for 5 min, decant the supernatant and gently resuspend in 0.2 ml of fresh 2× TY.

5 Add 25 µl of IPTG (25 mg/ml in water) and 25 µl of X-gal (20 mg/ml in DMF), mix and pour on to the surface of a pre-warmed LB-Amp plate. Spread over the agar surface using a sterile bent glass rod or sterile inoculating loop.

6 Allow 10–20 min for the liquid to diffuse into the agar, then invert and incubate overnight at 37°C.

### For pBR322, pAT153 or other nonlacZ-containing vectors

3 Add 1 ml of fresh 2× TY to the cells and incubate for 15–30 min at 37°C. Spread approximately 50 µl on lambda plates containing antibiotic using a sterile glass spreader. Incubate the plates overnight at 37°C.

*Protocol 14. Bacterial cell transformation*

# III  METHODS FOR DNA ISOLATION

## Methods available

### *Large-scale double-stranded DNA isolation* (see *Protocol 15*)

The method used for the isolation of large scale cosmid and plasmid DNA is an unpublished modification [3] of an alkaline lysis procedure [4,5] followed by equilibrium ultracentrifugation in cesium chloride–ethidium bromide gradients [1]. Briefly, cells containing the desired plasmid or cosmid are harvested by centrifugation, incubated in a lysozyme buffer, and treated with alkaline detergent. Detergent solubilized proteins and membranes are precipitated with sodium acetate, and the lysate is cleared first by filtration of precipitate through cheesecloth and then by centrifugation. The DNA-containing supernatant is transferred to a new tube, and the plasmid or cosmid DNA is precipitated by the addition of polyethylene glycol and collected by centrifugation. The DNA pellet is resuspended in a buffer containing cesium chloride and ethidium bromide, which is loaded into polyallomer tubes and subjected to ultracentrifugation overnight. The ethidium bromide stained plasmid or cosmid DNA bands, equilibrated within the cesium chloride density gradient after ultracentrifugation, are visualized under long wave UV light and the lower band is removed with a 5 cm$^3$ syringe. The intercalating ethid-

## References

1. Carter, J.M. and Milton, I.D. (1993) *Nucleic Acids Res.* **21**:1044.
2. Willis, E.H., Mardis, E.R., Jones, W.L. and Little, M.C. (1990) *BioTechniques*, **9**:92.
3. Pan, H., Chissoe, S. L., Bodenteich, A., Wang, Z., Iyer, K., Clifton, S.W., Crabtree, J.S. and Roe, B.A. (1994) *GATA* **11**:181.
4. *GS Gene Prep Manifold Instruction Manual.* BioRad Labs, Richmond, CA.
5. Messing, J. (1983) *Methods Enzymol.* **101**:20.
6. Sanger, F., Coulson, A.R., Barrell, B.G., Smith, A.J.H. and Roe, B.A. (1980) *J. Mol. Biol.* **143**:161.
7. Eperon, I.C. (1986) *Anal. Biochem.* **156**:406.
8. Mardis, E.R. and Roe, B.A. (1989) *BioTechniques*, **7**:736.
9. Chissoe, S.L., Wang, Y.F., Clifton, S.W., Ma, N., Sun, H.J., Lobsinger, J.S., Kenton, S.M., White, J.D. and Roe, B.A. (1991) *Methods: Companion Methods Enzymol.* **3**:55.
10. Federal Bureau of Investigation (1993) *RFLP Manual.* U.S. Government.
11. Sambrook, J., Fritsch, E.F. and Maniatis, T. (1989) *Molecular Cloning: A Laboratory Manual.* Cold Spring Harbor Laboratory Press, New York. Vol. 1,2,3.

ium bromide is separated from the DNA by loading the solution on to an equilibrated ion exchange column. The $A_{260}$ containing fractions are pooled, diluted, and ethanol precipitated, and the final DNA pellet is resuspended in buffer and assayed by restriction digestion and detected using agarose gel electorphoresis.

During the course of this work several modifications to the above protocol were made. For example, initially cell growth times included three successive overnight incubations, beginning with the initial inoculation of 3 ml of antibiotic containing media with the plasmid or cosmid-containing bacterial colony, and then increasing the culture volume to 50 ml, and then to 4 l. However, it was observed that recombinant cosmid DNA isolated from cell cultures grown under these conditions, in contrast to recombinant plasmid DNA, was contaminated with deleted cosmid DNA molecules. However, these deletions are avoided by performing each of the three successive incubations for 8 h instead of overnight, although a slight yield loss accompanied the reduced growth times.

Recently, a diatomaceous earth-based [6–9] method was used to isolate the plasmid or cosmid DNA from a cell lysate. The cell growth, lysis, and cleared lysate steps are performed as described above, but following DNA precipitation by polyethylene glycol, the DNA pellet is resuspended in RNase buffer and treated with RNase A and T1. Nuclease treatment is necessary to remove the RNA by digestion

## Protocols provided

15. *Large-scale double-stranded DNA isolation*
16. *Miniprep double-stranded DNA isolation*
17. *Large-scale M13 RF isolation*
18. *Single-stranded M13 DNA isolation using phenol*
19. *Biomek-automated modified Eperon isolation procedure for single-stranded M13 DNA*
20. *96-Well double-stranded template isolation*
21. *Genomic DNA isolation from blood*

*Methods for DNA isolation*

since RNA competes with the DNA for binding to the diatomaceous earth. After RNase treatment, the DNA containing supernatant is bound to diatomaceous earth in a chaotropic buffer of guanidine hydrochloride by incubation at room temperature. The DNA-associated diatomaceous earth is then collected by centrifugation, washed several times with ethanol buffer and acetate, dried, and then resuspended in buffer. The DNA is eluted during incubation at 65°C, and the DNA-containing supernatant is collected after centrifugation and separation of the diatomaceous earth particles. The DNA recovery is measured by taking absorbance readings at 260 nm. After concentration by ethanol precipitation, the DNA is assayed by restriction digestion.

### Miniprep double-stranded DNA isolation (see Protocol 16)

The standard method for the miniprep isolation of plasmid DNA includes the same general strategy as the large-scale isolation. However, smaller aliquots of antibiotic-containing liquid media inoculated with plasmid-containing cell colonies are incubated in a 37°C shaker for 12–16 h. After collecting the plasmid-containing cells by centrifugation, the cell pellet is resuspended in Tris-EDTA buffer. The cells are incubated successively with an RNase lysis buffer, alkaline detergent and sodium acetate. The lysate is cleared of precipitated proteins and membranes by centrifugation, and the plasmid DNA is recovered from the supernatant by ethanol precipitation.

The DNA is checked for concentration and purity using agarose gel electrophoresis against known standards. A typical yield for this method of DNA isolation is 10–15 µg of plasmid DNA from a 6 ml starting culture.

Since highly supercoiled DNA is desired for double-stranded DNA sequencing, a modification of this method employing diatomaceous earth [4–7] is sometimes used for isolation of double-stranded templates for DNA sequencing with fluorescent primers. After removal of the precipitated proteins and membranes, the plasmid-containing supernatant is incubated with diatomaceous earth and guanidine hydrochloride and this mixture is added into one of the 24 wells in the BioRad Gene Prep Manifold. The supernatant is removed by vacuum filtration over a nitrocellulose filter. The DNA-associated diatomaceous earth is washed to remove the guanidine hydrochloride with an ethanol buffer, and then dried by filtration. Elution buffer is added to the wells, and the DNA-containing solution is then separated from the diatomaceous earth particles by filtration into a collection tube. The collected DNA is concentrated by ethanol precipitation and simply assayed for concentration and purity by agarose gel electrophoresis against known standards. The approximate yield of double-stranded DNA is 3–5 µg of DNA from 6 ml of starting culture.

*Note* This is a typical miniprep until step 7, where for standard alkaline lysis purification the template is either precipitated and used for *Taq* cycle sequencing with the dye-labeled primers, or purified further via diatomaceous earth for *Taq* dye-labeled terminator cycle sequencing reactions. For Sequenase dye-labeled terminator sequencing reactions, the maxiprep procedure can be scaled down to isolate the sequencing template from 50–100 ml of bacterial culture.

### *Large-scale M13 RF isolation [5]* (see *Protocol 17*)
Double-stranded M13 RF is isolated for use in M13 *Sma*I-cut, dephosphorylated vector preparation, described below. The growth conditions of M13-infected bacterial cells (see *Figure 2*) appear convoluted, but result in a maximal amount of M13 replicative form (RF) molecules per cell. After the M13 RF-containing bacterial cells are harvested by centrifugation, the double-stranded molecules are isolated using the cesium chloride method for large-scale plasmid isolation, as described above. Briefly this entails alkaline cell lysis, sodium acetate precipitation of detergent-solubilized proteins and membranes, PEG DNA precipitation and extraction of ethidium bromide-stained DNA from a cesium chloride gradient after ultra-centrifugation. After removal of the ethidium bromide on an ion-exchange column, the DNA-containing fractions are detected by $A_{260}$ measurement and pooled, and the DNA is concentrated by ethanol

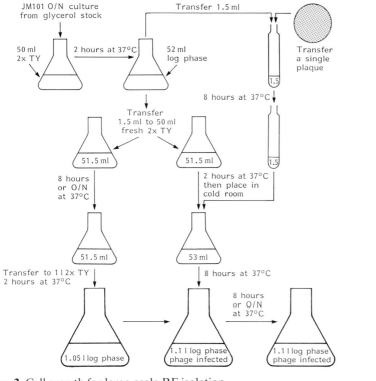

**Figure 2.** Cell growth for large-scale RF isolation.

*Methods for DNA isolation*

precipitation and assayed by restriction enzyme digestion and agarose gel electrophoresis.

**Single-stranded M13 DNA isolation using phenol** (see *Protocol 18*)
This isolation procedure [6] is the method of choice for preparation of M13-based templates to be used in Sequenase™ catalyzed dye-terminator reactions. A pre-incubated early log phase JM101 culture is prepared by transferring a thawed glycerol stock into 50 ml of liquid media and incubating for 1 h at 37°C with no shaking. M13 plaques are picked with a sterile toothpick and placed into 1.5 ml aliquots of the early log phase JM101 culture, which are incubated in a 37°C shaker for 4–6 h. After incubation, the bacterial cells are pelleted by centrifugation and the virus-containing supernatant is transferred to a clean tube. The phage particles are precipitated with PEG, collected by centrifugation and the pellet is resuspended in buffer. The phage protein coat is denatured and removed by one phenol and two ether extractions. After ethanol precipitation, the dried DNA pellet is resuspended in buffer, and the concentration and purity are evaluated by agarose gel electrophoresis against known standards.

**Biomek-automated modified Eperon isolation procedure**
(see *Protocol 19*)
This semi-automated method is a modification of a previously

reported procedure [7, 8], and allows simultaneous isolation of 48 single-stranded DNAs per Biomek 1000 robotic workstation within 3 h [9]. Basically, M13 plaques are picked with sterile toothpicks into aliquots of early log phase JM101, prepared as discussed above. The phage-infected cultures are incubated in a 37°C shaker for 4–6 h, transferred into microcentrifuge tubes, centrifuged to separate bacterial cells from the viral supernatant, and then carefully placed on the Biomek tablet. For each sample, two 250 μl aliquots are distributed robotically into two wells of a 96-well microtiter plate, and this process is repeated for each of the 48 samples until the entire 96 wells are filled. A solution of PEG then is added robotically to each well and mixed. The microtiter plate is covered with an acetate plate sealer, incubated at room temperature to precipitate the phage particles, and then centrifuged. The supernatant then is removed by inverting the plate and gently draining on a paper towel, without dislodging the pellet. After placing the microtiter plate back on the Biomek, a more dilute PEG solution is added robotically to each well. The plate is then covered with another sealer and centrifuged again. This rinse step aids in the removal of contaminating proteins and RNA. After removing the supernatant as before and placing the microtiter plate back on the Biomek, a Triton X-100 detergent solution is added robotically to each well. The plate is agitated gently and the sample from each pair of wells is transferred robotically to microcentrifuge tubes, which are then capped and placed in an 80°C water bath for 10 min to aid in the detergent solubilization of phage coat proteins. After a brief centrifugation to collect condensation, the single-stranded DNA is ethanol precipitated, dried and resuspended. An aliquot from each DNA sample is subjected to agarose gel electrophoresis to assay concentration and purity. The yield of single-stranded template is approximately 2–3 μg per sample.

### 96-Well double-stranded template isolation (see *Protocol 20*)

A manual as well as an automated procedure is available. The automated method is a modification of a previously reported procedure [11] which allows simultaneous isolation of 96 double-stranded DNAs per Biomek 1000 Automated Laboratory Workstation within 2 h. Basically, colonies containing double-stranded plasmids are picked with sterile toothpicks into media and incubated at 37°C for 24 h with shaking at 350 r.p.m. These cells are harvested by centrifugation and the pellets are resuspended either manually or robotically by the addition of TE-RNase solution. An alkaline lysis solution is used to lyse the cells and the lysate is precipitated with potassium acetate. The lysate is cleared by filtration and further concentrated by ethanol precipitation. An aliquot from each DNA sample is subjected to agarose gel eletrophoresis to assay concentration and purity. The yield of double-stranded template is approximately 3 μg per sample.

***Genomic DNA isolation from blood*** (see *Protocol 21*)

Genomic DNA isolation is performed according to the FBI protocol [10]. After the blood samples (stored at –70°C in EDTA Vacutainer™ tubes) are thawed, standard citrate buffer is added, mixed and the tubes are centrifuged. The top portion of the supernatant is discarded and additional buffer is added, mixed, and again the tube is centrifuged. After the supernatant is discarded, the pellet is resuspended in a solution of SDS detergent and proteinase K, and the mixture is incubated at 55°C for 1 h. The sample then is phenol extracted once with a phenol:chloroform:isoamyl alcohol solution and, after centrifugation, the aqueous layer is removed to a fresh microcentrifuge tube. The DNA is ethanol precipitated, resuspended in buffer and then ethanol precipitated again. After the pellet is dried, buffer is added and the DNA is resuspended by incubation at 55°C overnight; the genomic DNA solution is assayed by the polymerase chain reaction.

*Methods for DNA isolation*

# Protocol 15. Large-scale double-stranded DNA isolation

## Reagents

Acetone
Alkaline lysis solution (0.2 M NaOH, 1% SDS)
0.5% Ampicillin (Amp): add to media for final concentration of
    100 μg/ml
Cesium chloride high purity grade <99.99%
Diatomaceous earth (100 mg/ml)
Diatomaceous earth wash buffer
DNase-free RNase A
Dowex AG Resin (BioRad)
0.5 M EDTA, pH 8.0
100% Ethanol (McCormick Distilling Co., Inc.)
5 mg/ml Ethidium bromide (EtBr)
GET/lysozyme solution
6 M Guanidine HCl, pH 6.4 50 mM Tris-HCl, 20 mM EDTA
Isopropanol
LB Medium
1.0 M NaOH
10 M NaOH
50% PEG (8000), 0.5 M NaCl
RNase T1
10% SDS

## Equipment

Agarose gel electrophoresis apparatus
Autoclave
250 and 500 ml Bottles
Cheesecloth
Clean beaker
35 ml Corex glass tubes
1.5 ml Dowex column
Erlenmeyer flask
12× 75 mm Falcon tubes
100 ml Graduated cylinder
High speed centrifuge with large volume fixed-angle rotors
Ice-water bath
Long-wave UV light source
35 ml Polyallomer centrifuge tubes
Rubber stoppers
Spectrophotometer
Sterile toothpicks
5 ml Syringe with 25-gauge needle
Ultracentrifuge with fixed-angle rotor
Vacuum oven
37°C water bath

3 M Sodium acetate, pH 4.5
TE 10:01 buffer
TE 10:1 buffer
TE 100:10 buffer
1.0 M Tris–HCL, pH 7.6
1.0 M Tris–HCl, pH 8.0

## Procedure

1  Pick a colony of bacteria harboring the plasmid DNA of interest into a 12× 75 mm Falcon tube containing 2 ml of LB media supplemented with the appropriate antibiotic (typically ampicillin at 100 µg/ml) and incubate at 37°C 8–10 h with shaking at 250 r.p.m. Transfer the culture to an Erlenmeyer flask containing 50 ml of similar media, and incubate further for 8–10 h. Transfer 12.5 ml of the culture to each of 4 l of similar media, and incubate for an additional 8–10 h.

2  Harvest the cells by centrifugation at 10 000 $g$ for 20 minutes in 500 ml bottles in the high speed centrifuge using a fixed-angle rotor. Re-suspend the cell pellets in old media and transfer to two bottles, centrifuge as before and decant the media. [1]

3  Resuspend the cell pellets in a total of 70 m of GET/lysozyme solution (35 ml for each bottle) by gently loosening the pellet with a spatula and incubate for 10 minutes at room temperature. [1]

## Notes

This procedure takes about 6 hours.

[1] Do not vortex the lysate at any time because this may shear the chromosomal DNA.

*Protocol 15.  Large-scale double-stranded DNA isolation*

4 Add a total of 140 ml of alkaline lysis solution (70 ml for each bottle), gently mix, and incubate for 5 min in an ice-water bath.

5 Add 105 ml of 3 M sodium acetate, pH 4.5 (52.5 ml for each bottle), cap tightly, mix gently by inverting the bottle a few times, and incubate in an ice-water bath for 30–60 minutes.

6 Clear the lysate of precipitated SDS, proteins, membranes and chromosomal DNA by pouring through a double layer of cheesecloth. Transfer the lysate into a 250 ml centrifuge bottle, centrifuge at 15 000 $g$ for 30 min at 4°C in a high speed centrifuge using the fixed-angle rotor.

### For cesium chloride gradient purification

7 Pool the cleared supernatants into a clean beaker, add one-fourth volume of 50% PEG/0.5 M NaCl, swirl to mix, and incubate in an ice-water bath for 1–2 hours.

8 Collect the PEG-precipitated DNA by centrifugation in 250-ml bottles at 7000 r.p.m. for 20 min at 4°C in a high speed centrifuge using the fixed-angle rotor.

9 Dissolve the pellets in a combined total of 32 ml of TE buffer (100:10), 10 ml of 15 ml of 5 mg/ml ethidium bromide, and 37 g of cesium chloride (final concentration of cesium chloride should be 1 g/ml).

10 Transfer the sample into 35 ml polyallomer centrifuge tubes, remove air bubbles, seal with rubber stoppers, and crimp properly.

11 Centrifuge at 150 000 $g$ for 16–20 h at 15–20°C in an ultracentrifuge using a fixed angle rotor.

12  Visualize the ethidium bromide stained DNA under long-wave UV light, and remove the lower DNA band using a 5 ml syringe and a 25 gauge needle. (It may be helpful first to remove and discard the upper band.)

13  To remove the ethidium bromide, load the DNA sample on to an equilibrated 1.5 ml Dowex column, and collect 0.5 ml fractions. Equilibrate the Dowex AG resin by successive centrifugation, resuspension, and decanting with 1 M NaOH, water, and then 1 M Tris-HCl, pH 7.6 until the Dowex solution has a pH of 7.6.

14  Pool fractions with an $A_{260}$ of 1.00 or greater into 35 ml Corex glass tubes, add 1 volume of double-distilled water, and ethanol precipitate by adding 2.5 vol of cold 95% ethanol. Incubate for at least 2 h at $-20°C$, centrifuge at 10 000 $g$ for 45 min in a high speed centrifuge using a fixed-angle rotor. Gently decant the supernatant, add 70% ethanol, centrifuge as before, decant, and dry the DNA pellet in a vacuum oven.

15  Resuspend the DNA in TE (10:0.1) buffer.

### For diatomaceous earth-based purification

7  Pool the supernatants from Step 6 into 500 ml bottles and add DNase-free DNase A and RNase T1 such that the final concentration of RNase A is 40 μg/ml and RNase T1 is 40 U/ml. Incubate in a 37°C for 30 min.

8  Add an equal volume of isopropanol and precipitate at room temperature for 5 min. Centrifuge at 14 000 $g$ for 30 min in a high speed centrifuge using a fixed-angle rotor. Decant the supernatant and drain the DNA pellet.

9  Resuspend each DNA pellet in 20 ml of TE (10:1) buffer, and add 40 ml

*Protocol 15. Large-scale double-stranded DNA isolation*

of defined diatomaceous earth in guanidine-HCl (20 mg/ml, diluted from 10 mg/ml stock) to each bottle. Allow the DNA to bind at room temperature for 5 min with occasional mixing. Centrifuge at 14 000 $g$ for 10 min in a high speed centrifuge using a fixed-angle rotor.

10 Decant the supernatant, resuspend each pellet in 40 ml of diatomaceous earth wash buffer, and centrifuge as above.

11 Decant the supernatant, resuspend each pellet in 40 ml of acetone, and centrifuge as above.

12 Decant the supernatant and dry the pellet in a vacuum oven.

13 Resuspend the pellet in 20 ml of TE (10:1) buffer, and elute the bound DNA by incubation at 65°C for 10 min with intermittent mixing.

14 Remove the diatomaceous earth by centrifugation at 14,000 $g$ for 10 min in a high speed centrifuge using a fixed-angle rotor. Repeat if necessary.

15 Combine the DNA-containing supernatants and precipitate the DNA in 35 ml Corex glass tubes adding 2.5 vols of cold 95% ethanol/acetate. Incubate at -20°C for 1 h, centrifuge at 9000 r.p.m. for 10 min. Decant the supernatant and wash with 1 vol of 70% ethanol, spin as before and decant the supernatant. Dry the DNA pellet in a vacuum dryer until dry.

16 Resuspend the dried DNA pellet in 2 ml of TE (10:0.1) buffer and assay for concentration by absorbance readings at 260 nm or by agarose gel electrophoresis.

## Pause points

1 The cell pellets can be frozen at −70°C at this point.

## Reagents

Alkaline lysis solution (see Appendix A)

0.5% Ampicillin (Amp): add to media for final concentration of 100 µg/ml.

Diatomaceous earth (100 mg/ml)

Diatomaceous earth wash buffer (see Appendix A)

0.5 M EDTA, pH 8.0

95% Ethanol: 95 ml of 100% ethanol, add 5 ml of water⚠

100% Ethanol⚠

6 M Guanidine-HCl, pH 6.4, 50 mM Tris-HCl, pH 7.6, 20 mM EDTA pH 8.0

Isopropanol⚠

1.0 M NaOH⚠

10 M NaOH⚠

RNase A

RNase T1: RNase T1 (Sigma R-8251) (100 000 U/0.2 ml), diluted fourfold in 10 mM Tris–HCl pH 7.6

3 M Sodium acetate, pH 4.5

10% Sodium dodecylsulfate (SDS)

10× TB salts

TE-RNase A (40 µg/ml)

TE (10:0.1) buffer

TE (10:1) buffer:

Terrific Broth (TB)

TE-RNase solution

1.0 M Tris–HCl, pH 7.6

1.0 M Tris–HCl, pH 8.0

## Equipment

Agarose gel electrophoresis apparatus

17× 100 mm Falcon tube

100 ml Graduated cylinder

Ice-water bath

Low speed table-top centrifuge

1.5 ml Microcentrifuge tubes

Prep-A-Gene manifold (BioRad)

Prep-A-Gene nitrocellulose membrane (BioRad)

Repeating pipette

1.5 ml Screw-capped tubes

Sterile toothpicks

Vortex mixer

## Procedure

1 Pick, with sterile toothpicks, a colony of bacteria harboring the plasmid DNA of interest into a 17× 100 mm Falcon tube containing 6 ml of TB media supplemented with the appropriate antibiotic (typically ampicillin at 100 µg/ml) and incubate at 37°C for 16–18 h with shaking at 250 r.p.m.

2 Harvest the cells by centrifugation at 1500 $g$ for 5 min in a table-top centrifuge and decant the supernatant.[1]

3 Resuspend the cell pellets in 0.2 ml of TE-RNase solution [TE (50:10) buffer containing 40 µg/ml RNase A; some also add RNase T1 to a final concentration of 10 U/µl] by gentle vortexing, add 0.2 ml of alkaline lysis solution, gently mix and incubate for 15 min at room temperature.

4 Add 0.2 ml of 3 M sodium acetate, pH 4.5, mix gently by swirling, transfer to 1.5 ml microcentrifuge tubes and incubate in an ice-water bath for 15 min.

5 Clear the lysate of precipitated SDS, proteins, membranes and chromosomal DNA by centrifugation at 12 000 $g$ for 15 min in a microcentrifuge at 4°C.

6 Transfer the supernatant to a fresh 1.5 ml microcentrifuge tube, incubate in an ice-water bath for 15 min, centrifuge as above for an additional 15 min and transfer the supernatant to a clean 1.5 ml tube.

### For standard alkaline lysis purification

7 Precipitate the DNA by adding 1 ml of 95% ethanol, and resuspend the dried DNA pellet in 100–200 µl TE (10:0.1) buffer. Electrophorese an aliquot of the DNA sample on a 0.7% agarose gel to determine the concentration and purity (see *Protocol 4*).

## Notes

This procedure takes about 3 hours.

### For diatomaceous earth-based purification

7 Add 1 ml of defined diatomaceous earth in guanidine-HCl (20 mg/ml) and allow the DNA to bind at room temperature for 5 min with occasional mixing. Meanwhile soak the Prep-A-Gene nitrocellulose membrane in isopropanol for at least 3 min and assemble the Prep-A-Gene manifold as described in the manual.

8 Turn on the vacuum pump and adjust the vacuum level to 20 cm (8 in) mercury, let the membrane dry for 1 min and then release the vacuum.

9 Pour the well mixed samples into the wells of the Prep-A-Gene manifold and filter using the same vacuum until all the liquid is filtered through.

10 Wash the samples four times with 250 µl of diatomaceous earth wash buffer, using a repeat pipette, allowing all of the liquid to filter through between washes.

11 Reduce the vacuum to 12.5 cm (5 in) of mercury before turning the vacuum off at the stopcock. Without unscrewing the black clamps, release the white clamps and place the collection rack with clean 1.5 ml screw-capped tubes into the manifold. Clamp the manifold with the white clamps, and apply 300 µl of TE (10:1) buffer heated to 65°C and pull the eluted DNA through at 12.5 cm (5 in) of mercury. After the liquid has filtered through, raise the vacuum to 25–30 cm (25–30 in) of mercury, and let the membrane dry for 1 min.

12 Turn off the vacuum at the stopcock and remove the collection rack containing the tubes. Ethanol precipitate the DNA and resuspend the dried DNA pellet in 30 µl of TE (10:0.1) buffer.

## Pause points

[1] The cell pellets may be frozen at −20°C at this point.

## Reagents

Cesium chloride high purity grade (<99.99%)
DNase-free RNase A
0.5 M EDTA, pH 8.0
70% Ethanol⚠
95% Ethanol⚠
5 mg/ml Ethidium bromide⚠
Glycerol stock of JM101
Lysozyme solution
10 M NaOH⚠
3 M Sodium acetate, pH 4.5
1× STB buffer
TE (10:0.1) buffer
TE (100:10) buffer
1.0 M Tris–HCl, pH 7.6
1.0 M Tris–HCl, pH 8.0
50:2:10 TTE
2× TY medium

## Equipment

Autoclave
250 ml Centrifuge bottles
500 ml Centrifuge bottles
Cheesecloth
35 ml Corex glass tubes
Dowex AG column (Biorad)
Erlenmeyer flasks
High speed centrifuge with the 6× 250 ml fixed-angle rotor
Ice-water bath
Long-wave UV light
35 ml Polyallomer centrifuge tubes
Rubber stoppers
Ultracentrifuge and fixed-angle rotor
5 cm$^3$ Syringe with 25 gauge needle
Vacuum oven
37°C Water bath

## Procedure

1  Prepare an early log phase culture of JM101 by inoculating an Erlenmeyer flask containing 50 ml of 2× TY with a glycerol stock of

## Notes

This procedure takes approximately 24 hours.

JM101 and pre-incubating for 1 h in a 37°C water bath, with no shaking. Pick a plaque representing the desired M13 clone into four 1.5 ml aliquots of early log phase JM101, and incubate according to the procedure displayed in Figure 2 to give 4 l of M13-infected bacteria.

2   Harvest the cells by centrifugation at 10 000 $g$ for 20 min in 500 ml bottles in a high speed centrifuge using a fixed-angle rotor. Resuspend the cell pellets in fresh 2× TY media to remove contaminating extracellular phage, and transfer to two bottles, centrifuge as before and decant the media. $\boxed{1}$

3   Resuspend the cell pellets in a total of 120 ml (30 ml for each bottle) of 1× STB buffer by gently loosening the pellet with a spatula. Add a total of 24 ml of lysozyme solution (6 ml for each bottle), gently mix and incubate for 5 min in an ice-water bath.

4   Add 48 ml of 50:2:10 TTE buffer (12 ml for each bottle) and 2 ml of RNase A (10 mg/ml) (0.5 ml for each bottle), gently mix and incubate in an ice-water bath for 5 min.

5   Clear the lysate of precipitated proteins, membranes and chromosomal DNA by pouring through a double layer of cheesecloth. Transfer the lysate into 250 ml centrifuge bottles, centrifuge at 15 000 $g$ for 30 min at 4°C in a high speed centrifuge using a fixed-angle rotor.

6   Add 6 ml of 5 mg/ml ethidium bromide, and cesium chloride such that the final concentration of cesium chloride is 1 g/ml.

*Protocol 17. Large-scale M13 RF isolation*

7  Transfer the sample into 35 ml polyallomer centrifuge tubes and top off with a 1:1 solution of TE (100:10) buffer and cesium chloride, remove air bubbles, and seal the tube properly.

8  Centrifuge at 300 000 $g$ for 16–20 h at 15–20°C in an ultracentrifuge using a titanium fixed-angle rotor.

9  Visualize the ethidium bromide-stained DNA under long-wave UV light, and remove the lower DNA band using a 5 ml syringe and a 25 gauge needle. (It may be helpful to remove and discard the upper band first.)

10  To remove the ethidium bromide, load the DNA sample on to a 1.5 ml Dowex AG (BioRad) column, equilibrated with 1 M Tris–HCl, pH 7.5, and collect 0.5 ml fractions.

11  Pool fractions with an $A_{260}$ of 1.00 or greater into 35 ml Corex glass tubes, add 1 vol. of double-distilled water, and ethanol precipitate by adding 2.5 vol. of cold 95% ethanol. Incubate for at least 2 h at −20°C, centrifuge at 10 000 r.p.m. for 45 min in a high speed centrifuge using a fixed-angle rotor. Gently decant the supernatant, add 70% ethanol, centrifuge as before, decant and dry the DNA pellet in a vacuum oven.

12  Resuspend the DNA in TE (10:0.1) buffer.

## Pause points

[1] The cell pellets may be frozen at −70°C at this point.

# Protocol 18. Single-stranded M13 DNA isolation using phenol

## Reagents

0.5 M EDTA, pH 8.0
Phenol, TE (10:1)-saturated (store at 4°C)
0% Polyethylene glycol (PEG), 2.5 M NaCl
TE (10:0.1) buffer
1.0 M Tris–HCl, pH 7.6
Water-saturated ether

## Equipment

12× 75 mm Falcon tubes
Microcentrifuge
1.5 ml Microcentrifuge tubes
Sterile toothpicks
Vortex mixer

## Procedure

1  Prepare an early log phase culture of JM101, as in *Protocol 17*, and pick M13-based plaques with sterile toothpicks into 12× 75 mm Falcon tubes containing 1.5 ml aliquots of the cells. Incubate for 4–6 h at 37°C with shaking at 250 r.p.m.

2  Transfer the culture to 1.5 ml microcentrifuge tubes and centrifuge for 15 min at 12 000 $g$ at 4°C.

3  Pipette the top 1 ml of supernatant to a fresh 1.5 ml microcentrifuge tube containing 0.2 ml of 20% PEG, 2.5 M NaCl to precipitate the phage particles. Mix by inverting several times and incubate for 15–30 min at room temperature.

## Notes

This procedure takes 4 hours once the cell growth has been accomplished.

4   Centrifuge for 15 min at 12 000 $g$ at 4°C to collect the precipitated phage. Decant the supernatant and remove residual PEG supernatant by suctioning twice.

5   Resuspend the pellet in 100 µl of 10 mM Tris–HCl, pH 7.6 by vortexing, and add 50 µl of TE-saturated phenol.

6   Extract the DNA with phenol and twice with ether, as discussed in Protocol 1, and then ethanol precipitate.

7   Resuspend the dried DNA in 6 µl of TE (10:0.1) for use in single-stranded Sequenase™-catalyzed dye-terminator sequencing reactions.

*Protocol 18. Single-stranded M13 DNA isolation using phenol*

# Protocol 19. Biomek-automated modified Eperon isolation procedure

## Reagents

0.5 M EDTA, pH 8.0
95% Ethanol/0.12 M sodium acetate (ethanol/acetate)⚠
10 M NaOH⚠
PEG:TE rinse solution
Polyethylene glycol (PEG) 8000 (Fisher P156-3)
20% Polyethylene glycol (PEG) 8000, 2.5 M NaCl
TE (10:0.1) buffer
1.0 M Tris–HCl, pH 7.6
TTE
2× TY medium

## Equipment

Acetate plate sealers (Dynatech)
Autoclave
Biomek 1000 automated workstation
12× 75 mm Falcon tubes
Low speed table-top centrifuge able to spin microtiter plates
1.5 ml Microcentrifuge tubes
P250 tips
Paper tissues
Sterile toothpicks
96-Well flat-bottomed microtiter plate (Dynatech)

## Procedure

The entire procedure will require nine rows of P250 tips (counting from the center of the Biomek 1000 tablet towards the left) for the isolation of 48 templates (48ISOL). The reagent module should contain 20% PEG-8000, TTE and ethanol/acetate, respectively.

1  Prepare an early log phase JM101 culture in 50 ml of 2× TY, as in *Protocol 17* above.

2  Using sterile toothpicks, transfer individual M13 plaques into 12× 75 mm

## Notes

This procedure takes about 3 hours once bacterial growth has been accomplished.

① Growth for longer than 6 h results in cell lysis and contamination of the phage DNA by cellular proteins and nucleic acids.

Falcon tubes containing 1 ml of early log phase cell cultures, and incubate for 4–6 h at 37°C with shaking at 250 r.p.m. ①

3  Separate the bacterial cells from the viral-containing supernatant by centrifugation at 10 000 $g$ for 15 min at 4°C. Carefully open the tubes and place them on the Biomek tablet.

4  The Biomek will distribute two 250 µl aliquots of viral supernatant per sample into the wells of a 96-well flat-bottomed microtiter plate. The Biomek then will add 50 µl of 20% PEG/2.5 M NaCl solution to each well, and mix by pipetting up and down.

5  Cover the plate with an acetate plate sealer and incubate at room temperature for 15 min.

6  Pellet the precipitated phage by centrifuging the plate at 1000 $g$ (2400 r.p.m.) for 20 min in a table-top centrifuge. Remove the plate sealer and drain the PEG from the plate by gently placing it upside down on a tissue.

7  Return the plate to the tablet, and the Biomek will robotically add 200 µl of PEG:TE rinse solution to each well. Cover the plate with a plate sealer, centrifuge and drain, as above.

8  Return the plate to the tablet, and the Biomek will add 70 µl of TTE solution to each well. Remove and gently agitate to resuspend.

9  The Biomek then will pool the contents from each pair of wells robotically into 1.5 ml microcentrifuge tubes.

10  Incubate the tubes at 80°C for 10 min to denature the viral protein coat, and then centrifuge briefly to reclaim condensation.

11  Ethanol precipitate the DNA by adding 500 µl of ethanol/acetate to each tube, as described in *Protocol 2*.

12  Resuspend the DNA templates in 20 µl of TE (10:0.1) buffer.

*Protocol 19. Biomek-automated modified Eperon isolation procedure*

## Reagents

Alkaline lysis solution (see Appendix A)
0.5% Ampicillin (Amp) (add to media for final concentration of 100 μg/ml)
0.5 M EDTA, pH 8.0
100% Ethanol⚠
10 M NaOH⚠
3 M Sodium acetate, pH 4.5
10× TB salts
Terrific Broth (TB)
TE-RNase solution
TE (10:0.1) buffer
1.0 M Tris–HCl, pH 7.6

## Equipment

Beckman 96-well block
Biomek 1000 Automated Laboratory Workstation
12-Channel pipette
Dynatech 96-well flat bottomed microtiter plate
Low speed table-top centrifuge
Millipore filter plate
Paper towel
PolyFiltronics 96-well cover
QiaVac Vacuum Manifold 96
Robbins PCR reaction tube
Savant Speed-Vac
Shaker/incubator
Sterile toothpicks

## Procedure

### *Manual double-stranded isolation method*

The following is a manual, 96-well double-stranded sequencing template isolation procedure that has been developed in our laboratory. A similar procedure that has been automated on the Biomek is presented below.

1  Pick individual shotgun clones off a plate with a sterile toothpick and deposit each separately into a 96-well block containing 1.5 ml of TB

## Notes

This procedure takes about 4 hours.

①  Do not use a harsh filtration as the plates are fragile and will lose their seal.

media per well. Keep the toothpick in media for about 5 min to allow the cells to diffuse into the media, remove the toothpicks, cover the 96-well block with the loose fitting lid and allow the cells to grow for 24 h in the 37°C shaker/incubator at 350 r.p.m.

2  Remove the block from the shaker/incubator and collect the cells by centrifugation at 1000 $g$ for 7 min. [1]

3  Thaw cells, if frozen, and add 200 µl of TE-RNase A solution containing RNase T1, mix by pipetting up and down 4–5 times to resuspend the cell pellet and then incubate in the 37°C incubator/shaker for 5 min at 350 r.p.m. to mix more thoroughly.

4  Remove the block from the incubator/shaker and then add 200 µl of alkaline lysis solution. Shake the block by hand to mix the reagents and then incubate at room temperature for 1 h with intermittent swirling.

5  Add 200 µl of either 3 M potassium or sodium acetate, pH 4.5, and place the block in the 37°C shaker/incubator for 5 min at 350 r.p.m. to mix thoroughly and shear genomic DNA to reduce the viscosity of the solution. Place the block at –20°C for 30 min.

6  Centrifuge the block in the centrifuge at 1000 $g$ at 4°C for 30 min.

7  Carefully remove 400 µl of the supernatant from each well in the 96-well block with the 12-channel pipette and transfer to a second 96-well block, being careful not to transfer any cell debris. Transfer 10 µl of supernatant into the respective cycle sequencing reaction tubes, and precipitate with 30 µl of 95% ethanol (without added acetate). After storage at –20°C for 30 min, the pellet is collected by centrifugation, washed three times with

*Protocol 20.  96-Well double-stranded template isolation*

70% ethanol, and dried directly in the cycle sequencing reaction tubes. Prior to adding the fluorescent terminator cycle sequencing reaction mix, the dried templates should be stored at −20°C. An additional 75 μl of the supernatant is transfered to a Robbins PCR reaction tube (in 96-well tube format) and precipitated with 200 μl of 95% ethanol, washed three times with 70% ethanol and stored dry at −20°C for future use.

## *Automated double-stranded isolation method*

The following is an automated, 96-well double-stranded sequencing template isolation procedure that has been developed in our laboratory.

1 Pick colonies using a toothpick into a 96-well block containing 1.5 ml of TB with TB salts and the appropriate antibiotic, and shake for 24 h at 350 r.p.m. in a 96-well block with cover.

2 Harvest the cells by centrifugation at 1000 *g* for 7 min. Pour off the supernatant and allow the pellets to drain inverted.⊡1

3 Turn on Biomek, begin the program DSISOL2 (available directly from the authors) and set up the Biomek as indicated in the configuration function on the screen. Specifically, you should put TE-RNase solution in the first module, alkaline lysis solution in the second reagent module and 3 M potassium acetate, pH 4.5 in the third module.

4 Place the 96-well block containing cells on to the Biomek tablet at the position labeled '1.0 ml Minitubes'. Place a Millipore filter plate in the position labeled '96-well flat bottomed microtiter plate'.

5 Press ENTER to continue with the program.

6 First the Biomek will add 100 µl of TE-RNase solution to the cell pellets and mix to partially resuspend.

7 Next, the Biomek will add 100 µl of alkaline lysis solution to the wells of the filter plate.

8 The Biomek then will mix the cell suspension again, transfer the entire volume to the filter plate containing alkaline lysis solution, and mix again. Set up the filtration apparatus with a clean 96-well block to collect the filtrate (wash and re-use the block used for growth).

9 The Biomek will add 100 µl of 3 M potassium acetate, pH 4.5 to the wells of the filter plate and mix at the sides of the wells. Some choose to place the filter plate at –20°C for 5 min at this point. Transfer the filter plate to the QiaVac Vacuum Manifold 96 and filter using water vacuum only.① This will typically take less than 20 min.

10 The supernatant collected in the 96-well block is the crude DNA and must be ethanol precipitated before use by the addition of 1 ml of 100% ethanol and incubation at –20°C for at least 30 min.

11 Centrifuge for 25 min at 1500 $g$ (3000 r.p.m.) in a refrigerated centrifuge.

12 Decant and wash with 500 µl of 70% ethanol and centrifuge for an additional 5 min at 1500 $g$ (3000 r.p.m.)

13 Decant the supernatant, drain inverted on a paper towel. Dry under vacuum.

14 Resuspend in 50 µl of TE (10:0.1) buffer for use in dye primer or dye terminator sequencing chemistry.

## Pause points

1 The cells may be stored frozen at –20°C in the block at this stage.

## Reagents

0.5 M EDTA, pH 8.0
70% Ethanol⚠
100% Ethanol⚠
10 M NaOH⚠
Phenol:chloroform:isoamyl alcohol (25:25:1)⚠
Proteinase K (Sigma): 20 mg/ml in water
2 M Sodium acetate
10% Sodium dodecylsulfate (SDS)
1× Standard saline-citrate (SSC)
20× Standard saline-citrate (SSC)

TE (10:1) buffer
TE-saturated phenol (store at 4°C)⚠
1.0 M Tris–HCl, pH 7.6

## Equipment

EDTA Vacutainer™ tubes
Liquid blood samples
Microcentrifuge
1.5 ml Microcentrifuge tubes
Savant Speed-Vac
Vortex mixer

## Procedure

1  Obtain the liquid blood samples in EDTA Vacutainer™ tubes frozen at
   −70°C.①

2  Thaw the frozen samples, add 0.8 ml of 1× SSC buffer and mix.
   Centrifuge for 1 min at 12 000 $g$ in a microcentrifuge.

3  Remove 1 ml of the supernatant and discard into disinfectant.

4  Add 1 ml of 1× SSC buffer, vortex, centrifuge as above for 1 min and
   remove all of the supernatant.

## Notes

This procedure takes about 2 hours.

①  Treat all blood samples as biohazards.

5 Add 375 µl of 0.2 M sodium acetate to each pellet and vortex briefly. Then add 25 µl of 10% SDS and 5 µl of proteinase K, vortex briefly and incubate for 1 h at 55°C.

6 Add 120 µl of phenol:chloroform:isoamyl alcohol and vortex for 30 sec. Centrifuge the sample for 2 min at 12 000 $g$ in a microcentrifuge tube.

7 Carefully remove the aqueous layer to a new 1.5 ml microcentrifuge tube, add 1 ml of cold 100% ethanol, mix and incubate for 15 min at −20°C.

8 Centrifuge for 2 min at 12 000 $g$ in a microcentrifuge. Decant the supernatant and drain.

9 Add 180 µl of TE (10:1) buffer, vortex and incubate at 55°C for 10 min.

10 Add 20 µl of 2 M sodium acetate and mix. Add 500 µl of cold 100% ethanol, mix and centrifuge for 1 min at 12 000 $g$ in a microcentrifuge.

11 Decant the supernatant and rinse the pellet with 1 ml of 70% ethanol. Centrifuge for 1 min at 12 000 $g$ in a microcentrifuge.

12 Decant the supernatant, and dry the pellet in a Savant Speed-Vac for 10 min (or until dry).

13 Resuspend the pellet by adding 200 µl of TE (10.1) buffer. Incubate overnight at 55°C, vortexing periodically to dissolve the genomic DNA. Store the samples at −20°C.

*Protocol 21. Genomic DNA isolation from blood*

# IV METHODS FOR DNA SEQUENCING

## Methods available

Many laboratories, including ours, now have implemented fluorescent-based methods for DNA sequencing as the method of choice. However, we have included the following sections dealing with the methods for radiolabeled-based DNA sequencing for historical reasons and because the reader may find them useful.

### Bst-*catalyzed radiolabeled DNA sequencing* (see *Protocol 22*)

Among the enzymes used for radiolabeled DNA sequencing, *Bst* DNA polymerase has been the enzyme of choice in our laboratory. *Bst* DNA polymerase-catalyzed radiolabeled two-step sequencing reactions [1] are modified from those presented earlier [2] by altering the absolute amounts and the relative deoxy/dideoxynucleotide ratios in the termination mixes. Two separate termination mixes provide optimal overlap for sequence data starting in the polylinker and extending to approximately 600 bases from the priming site. This two-step format eliminates the need for the chase required in the *Bst* one-step reaction [2].

Each extension reaction contained 500–750 ng of Biomek-isolated single-stranded DNA, reaction buffer, nucleotide extension mix, oligonucleotide primer [typically M13 (–40) universal sequencing

## References

1. Chissoe, S.L., Wang, Y.F., Clifton, S.W., Ma, N., Sun, H.J., Lobsinger, J.S., Kenton, S.M., White, J.D. and Roe, B.A. (1991) *Methods: Companion Methods Enzymol.* **3**:55.
2. Mardis, E.R. and Roe, B.A. (1989) *BioTechniques*, **7**:736.
3. Siemenak, D.R., Sieu, L.C. and Slightom, J.L. (1991) *Anal. Biochem.* **192**:441.
4. Johnston-Dow, L., Mardis, E., Heiner, C. and Roe, B.A. (1987) *BioTechniques*, **5**:754.
5. Bodenteich, A., Chissoe, S., Wang, Y.F. and Roe, B.A. (1994) in *Automated DNA Sequencing and Analysis Techniques* (C. Venter, ed.), p. 42. Academic Press, London.
6. Protocol included with the ABI PRISM Sequenase™ Terminator Double-Stranded DNA Sequencing Kit, part number 401461.
7. Protocol included with the ABI PRISM Sequenase™ Terminator Single-Stranded DNA Sequencing Kit, part number 401458.
8. Lee, L.G., Connell, C.R., Woo, S.L., Cheng, R.D., McArdle, B.F., Fuller, C.W., Halloran, N.D. and Wilson, R.K. (1992) *Nucleic Acids Res.* **20**:2471.
9. Chen, E.Y. and Seeburg, P.H. (1985) *DNA*, **4**:165.

primer, see Appendix A), either [α-$^{32}$P]dATP or [α-$^{35}$S]dATP and *Bst* polymerase. After the reactions are extended for 2 min at 67°C and briefly centrifuged, four aliquots are removed and added to the appropriate base-specific termination mix. All nucleotide mixes contained the guanosine nucleotide analog, 7-deaza-dGTP (c$^7$dGTP), but differ in their deoxy/dideoxynucleotide ratios to yield fragments ranging in size from the beginning of the polylinker to greater than 300 bases from the primer, or fragments from about 150 to greater than 600 bases from the primer for 'short' or 'long' mixes, respectively. Following an incubation at 67°C for 10 min and a brief centrifugation, the reactions are stopped by the addition of loading dye, and incubated at 100°C. When desired, sequencing reactions are stored at −70°C prior to the addition of loading dye.

When double-stranded pUC-based subclones are used as templates, the amount of primer is doubled and a denaturing/annealing step is added. Here, 3 μg of plasmid DNA, isolated by the miniprep diatomaceous earth method (see *Protocol 16*), is mixed with primer, placed in a boiling-water bath, and rapidly cooled by plunging into an ethanol/dry-ice bath [3]. Following an incubation on ice, the remaining sequencing extension reagents (reaction buffer, nucleotide extension mix, either [α-$^{32}$P]dATP or [α-$^{35}$S]dATP, and *Bst* polymerase) are added. Reactions are performed as described above for single-stranded sequencing.

10. Olson, M., Hood, L., Cantor, C. and Botstein, D. (1989) *Science*, **245**:1434.
11. Khan, A.S., Wilcox, A.S., Hopkins, J.A. and Sikela, J.M. (1991) *Nucleic Acids Res.* **19**:1715.
12. Khan, A.S., Wilcox, A.S., Polymeropoulos, M.H., Hopkins, J.A., Stevens, T.J., Robinson, M., Orpano, A.K. and Sikela, J.M. (1992) *Nature Genetics*, **2**:180.
13. Mullis, K.B. and Faloona, F.A. (1987) *Methods Enzymol.* **155**:335.
14. Smith, L.M., Sanders, J.Z., Kaiser, R.J., Hughes, P., Dodd, C., Connell, C.R., Heiner, C., Kent, S.B.H. and Hood, L.E. (1986) *Nature*, **321**:674.
15. Barnes, W.M. (1987) *Methods Enzymol.* **152**:538.
16. Henikoff, S. (1984) *Gene*, **28**:351.

## Protocols provided

### Klenow fragment of DNA polymerase-catalyzed radiolabeled DNA sequencing (see *Protocol 23*)

Historically, the Klenow fragment of *E. coli* DNA polymerase has been the enzyme of choice for radiolabeled DNA sequencing. Many portions of the DNA sequencing protocol using this enzyme are similar to that described above for *Bst* polymerase but, because the Klenow fragment is not as heat stable as the *Bst* polymerase, care should be taken during the incubations not to exceed 55°C for the annealing step and 37°C for the combined elongation and termination steps. Briefly, an annealing mix containing buffer, primer and template DNA is incubated in the Klenow-based DNA sequencing reactions at 55°C for annealing for 5 min, followed by incubation for 30 min at 37°C for elongation. A dye/formamide/EDTA mixture then is added to each of the four separate reactions and, after concentration by incubation at 100°C for 10 min, the contents of each tube are loaded on to adjacent wells of a polyacrylamide gel to resolve the nested fragment set.

### Modified T7 DNA polymerase (Sequenase™)-catalyzed radiolabeled DNA sequencing (see *Protocol 24*)

Because a commercially available kit that includes the Sequenase™ enzyme is available, many laboratories presently use the modified T7 DNA polymerase-catalyzed reaction for routine DNA sequencing. The procedure employed is similar to that described above for the

Klenow fragment of *E. coli* DNA polymerase, but the resulting fragment set obtained with the Sequenase™ enzyme usually yields a much more uniform intensity in the resulting nested fragment set. Briefly, an annealing mix containing buffer, primer and template DNA is incubated in the Sequenase™-based DNA sequencing reactions at 55°C for annealing for 5 min, followed by incubation for 30 min at 37°C. Subsequently, a labeling mix containing the four deoxynucleotides (one of which is labeled in the alpha position with either $^{32}$P or $^{35}$S) is added and incubated for 5 min at room temperature. An aliquot of the annealing/labeling reaction is then added to each of four tubes containing the appropriate termination mix and incubated for an additional 5 min at 37°C. A dye/formamide/EDTA mixture is then added to each of the four separate reactions and, after concentration by incubation at 100°C for 10 min, the contents of each tube are loaded on to adjacent wells of a polyacrylamide gel to resolve the Sequenase™-generated nested fragment set.

### *Radiolabeled sequencing gel preparation, loading and electrophoresis* (see *Protocol 25*)

To prepare polyacrylamide gels for DNA sequencing [1, 4], the appropriate amount of urea is dissolved by heating in water and electrophoresis buffer, the respective amount of deionized acrylamide/bisacrylamide solution is added, and ammonium persulfate and TEMED are added to initiate polymerization. Immediately after the addition of the polymerizing agents, the gel solution is poured between two glass plates which have been taped together and separated by thin spacers corresponding to the desired thickness of the gel, taking care to avoid and eliminate air bubbles. Prior to taping, these glass plates are cleaned with Alconox detergent and hot water, rinsed with double distilled water and dried with a tissue. Typically, the notched glass plate is treated with a silanizing reagent and then rinsed with double-distilled water. After pouring, the gel is immediately laid horizontally, and a well-forming comb is inserted into the gel and held in place by metal clamps. The polyacrylamide gels are allowed to polymerize for at least 30 min prior to use. After polymerization, the comb and the tape at the bottom of the gel are removed. The vertical electrophoresis apparatus is assembled by clamping the top and bottom buffer wells on to the gel, and adding running buffer to the buffer chambers. The wells are cleaned by circulating buffer into the wells with a syringe and, immediately prior to the loading of each sample, the urea in each well is suctioned out with a flat-tipped loading pipette.

Each base-specific sequencing reaction terminated with the short termination mix is loaded using a flat-tipped loading pipette on to a 0.15 mm × 50 cm × 20 cm, denaturing 5% polyacrylamide gel and electrophoresed for 2.25 h at 22 mA. The reactions terminated with the long termination mix typically are divided in half and loaded on to two 0.15 mm × 70 cm × 20 cm denaturing 4% polyacrylamide gels. One gel is electrophoresed at 15 mA for 8–9 h and the other is electrophoresed for 20–24 h at 15 mA. After electrophoresis, the glass plates are separated and the gel is blotted on to Whatman paper, covered with plastic wrap, dried by heating on a Hoefer vacuum gel dryer and exposed to X-ray film. Depending on the intensity of the signal and whether the radiolabel is $^{32}P$ or $^{35}S$, exposure times vary from 4 h to several days. After exposure, the films are developed by processing in developer and fixer solutions, rinsed with water and air dried. The autoradiogram then is placed on a light box, the sequence is read manually and the data typed into a computer.

### Taq *polymerase-catalyzed cycle sequencing using fluorescent-labeled primers* (see *Protocol 26*)

Each base-specific fluorescent-labeled cycle sequencing reaction routinely includes approximately 100 or 200 ng of Biomek-isolated single-stranded DNA for A and C or G and T reactions, respectively. Double-stranded cycle sequencing reactions similarly contain approximately 200 or 400 ng of plasmid DNA, isolated using either the standard alkaline lysis or the diatomaceous earth-modified alkaline lysis procedures. All reagents except template DNA are added in one pipetting step from a pre-mix of previously aliquoted stock solutions stored at –20°C (see Appendix A). To prepare the reaction pre-mixes, reaction buffer is combined with the base-specific nucleotide mixes.

Prior to use, the base-specific reaction pre-mixes are thawed and combined with diluted *Taq* DNA polymerase and the individual fluorescent end-labeled universal primers (see Appendix A) to yield the final reaction mixes that are sufficient for 24 template samples.

Once the reaction mixes are prepared, four aliquots of single- or double-stranded DNA are pipetted into the bottom of each 0.2 ml thin-walled reaction tube, corresponding to the A, C, G and T reactions, and then an aliquot of the respective reaction mixes is added to the side of each tube. These tubes are part of a 96-tube/retainer set tray in a microtiter plate format, which fits into a Perkin Elmer Cetus Cycler 9600. Strip caps are sealed on to the tube/retainer set and the plate is centrifuged briefly. The plate is then placed in the cycler whose heat block had been pre-heated to 95°C, and the cycling program is started immediately. The cycling protocol consisted of 15–30 cycles of seven temperatures: 95°C denaturation, 55°C annealing, 72°C extension, 95°C denaturation 72°C extension, 95°C denaturation and 72°C extension, and is linked to a 4°C final soak file. At this stage, the reactions frequently are frozen and stored at −20°C for up to several days. Prior to pooling and precipitation, the plate is centrifuged briefly to recover the sample. The four reactions are then pooled into ethanol, and the DNA is precipitated and dried. These sequencing reactions can be stored for several days at −20°C. For further details see refs 1, 5.

***Fluorescent terminator reactions*** (see *Protocol 27*)

One of the major problems in DNA cycle sequencing is that when fluorescent primers [5] are used, the reaction conditions are such that the nested fragment set distribution is highly dependent upon the template concentration in the reaction mix. We have observed that the nested fragment set distribution for the DNA cycle sequencing reactions using the fluorescent-labeled terminators [6] is much less sensitive to DNA concentration than that obtained with the fluorescent-labeled primer reactions as described above. In addition, the fluorescent terminator reactions require only one reaction tube per template, while the fluorescent-labeled primer reactions require one reaction tube for each of the four terminators. This latter point allows the fluorescent-labeled terminator reactions to be pipetted easily in a 96-well format. The protocol used, as described, is easily interfaced with the 96-well template isolation and 96-well reaction clean-up procedures also described herein. By performing all three of these steps in a 96-well format, the overall procedure is highly reproducible and therefore less error prone.

***Sequenase™-catalyzed sequencing with dye-labeled terminators*** (see *Protocol 28*)

Single-stranded dye-terminator reactions required approximately 2 μg of phenol-extracted M13-based template DNA. The DNA is denatured and the primer annealed by incubating DNA, primer and buffer at 65°C. After the reaction has been cooled to room temperature, α-thio-deoxynucleotides, fluorescent-labeled dye-terminators and diluted Sequenase™ DNA polymerase are added and the mixture is incubated at 37°C. The reaction is stopped by adding ammonium acetate and ethanol, and the DNA fragments are precipitated and dried. To aid in the removal of unincorporated dye-terminators, the DNA pellet is rinsed twice with ethanol. The dried sequencing reactions can be stored up to several days at –20°C.

Double-stranded dye-terminator reactions require approximately 5 µg of diatomaceous earth-modified alkaline lysis purified plasmid DNA. The double-stranded DNA is denatured by incubating the DNA in NaOH solution at 65°C and, after incubation, primer is added and the reaction is neutralized by adding an acid buffer. Reaction buffer, α-thio-deoxynucleotides, fluorescent-labeled dye-terminators and diluted Sequenase™ DNA polymerase are then added and the reaction is incubated at 37°C. Ammonium acetate is added to stop the reaction and the DNA fragments similarly are precipitated, rinsed, dried and stored. The protocols are based on refs. 6, 7 and 8, and T. Favello and R.K. Wilson, personal communication.

### *Fluorescent DNA sequencing* (see *Protocol 29*)
Polyacrylamide gels for fluorescent DNA sequencing are prepared as described above, except that the gel mix is filtered prior to polymerization. Optically ground, low fluorescence glass plates are cleaned

carefully with hot water, distilled water and ethanol to remove potential fluorescent contaminants prior to taping. To aid in well formation, bind silane is applied to the well area. Denaturing 4.75% polyacrylamide gels are poured into 0.3 mm × 89 cm × 52 cm taped plates and fitted with 36-well-forming combs. After polymerization, the tape and the comb are removed from the gel and the outer surfaces of the glass plates are cleaned with hot water, and rinsed with distilled water and ethanol. The gel is assembled into an ABI sequencer, and then checked by laser-scanning. If baseline alterations are observed on the ABI-associated Macintosh computer display, the plates are recleaned. Subsequently, the buffer wells are attached, electrophoresis buffer is added and the gel is pre-electrophoresed for 10–30 min at 35 W.

Prior to sample loading, the pooled and dried reaction products are resuspended in formamide/EDTA loading buffer by vortexing and then heated at 90°C. A sample sheet is created within the ABI data collection software on the Macintosh computer which indicates the number of samples loaded and the mobility file to use for sequence data processing. After cleaning the sample wells with a syringe, the odd-numbered sequencing reactions are loaded into the respective wells using an automatic pipettor equipped with a flat gel loading tip. The gel then is electrophoresed for 5 min before the wells are cleaned again, and the even-numbered samples are loaded. The filter wheel used for dye-primers and dye-terminators is specified on the ABI 373A CPU, where electrophoresis conditions are also adjusted. Typically, electrophoresis and data collection are for 10 h at 35 W on the ABI 373A that is fitted with a heat-distributing aluminum plate in contact with the outer glass gel plate in the region between the laser stop and the sample loading wells [1].

After data collection, an image file is created by the ABI software which relates the fluorescent signal detected to the corresponding scan number. The software then determines the sample lane positions based on the signal intensities. After the lanes are tracked, the cross-section of data for each lane is extracted and processed by baseline subtraction, mobility calculation, spectral deconvolution and time correction. On the Macintosh computer, the collected data can be viewed in several formats. The overall graphics image of the gel can be displayed to assess the accuracy of lane tracking, and the data from each sample lane can be viewed as either a four-color raw fluorescent signal versus scan number, as a chromatogram of processed sequence data, or as a string of nucleotides. After processing, the sequence data files are transferred to a SPARCstation 2 using NFS Share.

### Double-stranded sequencing of cDNA clones containing long poly(A) tails using anchored poly(dT) primers (see *Protocol 30*)

Sequencing double-stranded DNA templates has become a common and efficient procedure [9] for rapidly obtaining sequence data while avoiding preparation of single-stranded DNA. Double-stranded templates of cDNAs containing long poly(A) tracts are difficult to sequence with vector primers which anneal downstream of the poly(A) tail. Sequencing with these primers results in a long poly(T) ladder followed by a sequence which is difficult to read. In an attempt to solve this problem, we synthesized three primers which contain $(dT)_{17}$ and either (dA), (dC) or (dG) at the 3′ end. We reasoned that the presence of these three bases at the 3′ end would 'anchor' the primers at the upstream end of the poly(A) tail and allow sequencing of the region immediately upstream of the poly(A) region.

Using this protocol, over 300 bp of readable sequence could be obtained. We have applied this approach to several other poly(A)-containing cDNA clones with similar results. Sequencing of the opposite strand of these cDNAs using insert-specific primers occurred directly upstream of the poly(A) region. The ability directly to obtain sequences immediately upstream from the poly(A) tail of cDNAs should be of particular importance to large-scale efforts to generate sequence-tagged sites (STSs) [10] from cDNAs [11, 12].

***cDNA sequencing based on PCR and random shotgun subcloning***
(see *Protocol 31*)
*Protocol 31* is a rapid and efficient method for sequencing cloned cDNAs based on PCR amplification [13], random shotgun cloning

(refs 1, 5, and S. Surzyski, personal communication) and automated fluorescent sequencing [14]. This method was developed in our laboratory because once the sequence of a genomic DNA-containing cosmid is obtained and putative exons are predicted, the corresponding cDNAs should be sequenced in a timely manner. However, the presently implemented directed cDNA sequencing strategies, that is primer walking [15] and exonuclease III deletion [16], are both time consuming and labor intensive, while the alternative, that is randomly shearing the intact plasmid followed by shotgun sequencing (refs 1, 5, and S. Surzyski, personal communication), leads to a significant number of clones containing the original cDNA cloning vector rather than the desired cDNA insert.

This is a polymerase chain reaction (PCR)-based approach where the 'universal' forward and/or reverse priming sites were excluded from the resulting PCR product by choosing a primer pair that lay between the usual 'universal' forward and reverse priming sites and the multiple cloning sites of the Stratagene Bluescript vector. These two PCR primers, with the sequence 5′-TCGAGGTCGACG-GTATCG-3′ for the forward or −16nts primer and 5′-GCCGCTC-TAGAACTAG TG-3′ for the reverse or +19nts primer, now have been used to amplify sufficient quantities of cDNA inserts in the 1.2–3.4 kb size range, so that the random shotgun sequencing approach described below could be implemented.

*Methods for DNA sequencing*

# Protocol 22. *Bst*-catalyzed radiolabeled DNA sequencing

## Reagents

1 mg/ml Bovine serum albumin (BSA) (aliquot and store at −20°C)

*Bst* dilution buffer

*Bst* 'long' termination 'A' mix:  8 μl of 0.5 mM dATP, 16.4 μl of 5 mM dCTP, 16.4 μl of 5 mM $c^7$dGTP, 16.4 μl of 5 mM dTTP, 11 μl of 5 mM ddATP, 50 μl of TE (50:1) buffer, 381.8 μl of sterile double-distilled water, aliquot (18 μl for 6 reactions) and store at −70°C

*Bst* 'long' termination 'C' mix:  16.4 μl of 5 mM dATP, 8 μl of 0.5 mM dCTP, 16.4 μl of 5 mM $c^7$dGTP, 16.4 μl of 5 mM dTTP, 6.5 μl of 5 mM ddCTP, 50 μl of TE (50:1) buffer, 386.3 μl of sterile double-distilled water, aliquot (18 μl for 6 reactions) and store at −70°C

*Bst* 'long' termination 'G' mix:  16.4 μl of 5 mM dATP, 16.4 μl of 5 mM dCTP, 8 μl of 0.5 mM $c^7$dGTP, 16.4 μl of 5 mM dTTP, 7 μl of 5 mM ddGTP, 50 μl of TE (50:1) buffer, 385.8 μl of sterile double-distilled water, aliquot (18 μl for 6 reactions) and store at −70°C

*Bst* 'long' termination 'T' mix:  16.4 μl of 5 mM dATP, 16.4 μl of 5 mM dCTP, 16.4 μl of 5 mM $c^7$dGTP, 8 μl of 0.5 mM dTTP, 15 μl of 5 mM ddTTP, 50 μl of TE (50:1) buffer, 377.8 μl of sterile double-distilled water, aliquot (18 μl for 6 reactions) and store at −70°C

*Bst* nucleotide extension mix:  3 μl of 5 mM dCTP, 3 μl of 5 mM $c^7$dGTP, 3 μl of 5 mM dTTP, 100 μl of buffer, 891 μl of sterile double-distilled water

*Bst* reaction buffer

*Bst* 'short' termination 'A' mix:  8 μl of 0.5 mM dATP, 16.4 μl of 5 mM dCTP, 16.4 μl of 5 mM $c^7$dGTP, 16.4 μl of 5 mM dTTP, 66 μl of 5 mM ddATP, 50 μl of TE (50:1) E buffer, 326.8 μl sterile double-distilled water, aliquot (18 μl for 6 reactions) and store at −70°C

*Bst* 'short' termination 'C' mix:  16.4 μl of 5 mM dATP, 8 μl of 0.5 mM dCTP, 16.4 of μl 5 mM $c^7$dGTP, 16.4 of μl 5 mM dTTP, 40 μl of 5 mM ddCTP, 50 μl of TE (50:1) buffer, 352.8 μl of sterile double-distilled water, aliquot (18 μl for 6 reactions) and store at −70°C

*Bst* 'short' termination 'G' mix:  16.4 μl of 5 mM dATP, 16.4 μl of 5 mM dCTP, 8 μl of 0.5 mM $c^7$dGTP, 16.4 μl of 5 mM dTTP, 54 μl of 5 mM ddGTP, 50 μl of TE (50:1) buffer, 338.8 μl of sterile double-distilled water, aliquot (18 μl for 6 reactions) and store at −70°C

*Bst* 'short' termination 'T' mix: 16.4 µl of 5 mM dATP, 16.4 µl of 5 mM dCTP, 16.4 µl of 5 mM c$^7$dGTP, 8 µl of 0.5 mM dTTP, 60 µl of 5 mM ddTTP, 50 µl of TE (50:1) buffer, 332.8 µl of sterile double-distilled water, aliquot (18 µl for 6 reactions) and store at −70°C

[α-$^{32}$P]dATP (110 Tbq/mmol; 3000Ci/mmol; PB 10384, Amersham) or [α-$^{35}$S]dATP (50TBq/mmol; 1500Ci/mmol; SJ 1304, Amersham)△

Diluted *Bst* polymerase (0.1 U/µl): dilute the *Bst* polymerase (BioRad 170-3406) in *Bst* dilution buffer

1.0 M Dithiothreitol (DTT) (aliquot and store at −20°C)

0.5 M EDTA, pH 8.0

1.0 M HEPES, pH 7.5

1.0 M MgCl$_2$

10 M NaOH△

Oligonucleotide primer (2.5 ng/µl)

PAGE gel loading dye: 0.03 g xylene cyanol FF, 0.03 g bromphenol blue, 0.2 ml 0.5 M EDTA, pH 8.0, de-ionized formamide to 10 ml

TE (50:1) buffer

1.0 M Tris–HCl, pH 7.6

1.0 M Tris–HCl, pH 8.5

## Equipment

Boiling-water bath

Ethanol/dry-ice bath△

Ice-water bath

Microcentrifuge

Microcentrifuge tubes

V-bottomed microtiter plate (Dynatech)

## Procedure

### For single-stranded DNA sequencing

1 Prepare the following extension reaction in a microcentrifuge tube:
- 750 ng of M13 template DNA
- 2 µl of *Bst* reaction buffer
- 2 µl of *Bst* nucleotide extension mix
- 1 µl of oligonucleotide primer (2.5 ng/µl)

## Notes

This procedure takes about 0.5 hours.

*Protocol 22. Bst-catalyzed radiolabeled DNA sequencing*

- 0.5–1 µl of [α-$^{32}$P]dATP or [α-$^{35}$S]dATP
- 1 µl of diluted *Bst* polymerase (0.1 U/µl)
- Sterile double-distilled water to 12 µl.

2 Incubate the reactions for 2 min at 67°C, and briefly centrifuge to reclaim condensation.

3 Remove 2.5 µl aliquots for each reaction into the four base-specific termination mixes (either short or long), already pipetted into a V-bottomed microtiter plate.

4 Incubate the reactions for 10 min at 67°C, and briefly centrifuge to reclaim condensation. [1]

5 Stop the reactions by the addition of 4 µl of loading dye and incubate for 5–7 min at 100°C.

### For double-stranded DNA sequencing

1 To denature the DNA and anneal the primer, incubate the following reagents in a boiling-water bath for 4–5 min and rapidly cool the reaction by plunging into an ethanol/dry-ice bath.
- 3 µg of plasmid DNA
- 5 ng of oligonucleotide primer
- Sterile double-distilled water to 9 µl.

2 Incubate the reaction in an ice-water bath for 5 min, and then add the following reagents:
- 2 µl of *Bst* reaction buffer
- 2 µl of *Bst* nucleotide extension mix
- 0.5–1 µl of [α-$^{32}$P]dATP or [α-$^{35}$S]dATP

- 1 µl of diluted *Bst* polymerase (0.1 U/µl)
- Sterile double-distilled water to 15 µl.

3 Proceed with the sequencing reaction as described above in Steps 2–5 for single-stranded templates.

## Pause points

1 It is possible to store the reactions at –70°C at this point.

*Protocol 22.  Bst-catalyzed radiolabeled DNA sequencing*

# Protocol 23. **Klenow-catalyzed radiolabeled DNA sequencing**

## Reagents

0.1 M Dithiothreitol (DTT), store at −20°C
DNA polymerase
Dye/formamide/EDTA solution
10× Hind buffer
5× Hind/DTT buffer
Klenow dilution buffer (KDB)

Klenow fragment of *E. coli*
Klenow labeling mix
Klenow termination mixes

## Equipment

Microcentrifuge
Water bath

## Procedure

1  Set up an annealing reaction for each clone to be sequenced:
   - 2.0 µl of distilled water
   - 2.0 µl of 5× Hind/DTT buffer
   - 1.5 µl of primer (2.5 ng/µl)
   - 5.0 µl of DNA (0.2 µg/µl)
   to give a total volume of 10.5 µl.

2  Incubate the annealing reactions at 55°C for 5 min, and then at 37°C for 30 min. During the annealing incubations prepare an 8 M urea, 6% poly-acrylamide gel (38× 50 cm) as described in *Protocol 25*, and set up four 0.5 ml microcentrifuge tubes per annealing reaction labeled as follows: A1, C1, G1, T1, etc.

## Notes

This procedure takes about 1 hour.

3  To each annealing reaction, add 2.0 μl of 1× Klenow labeling mix, 1.5 μl of [α-$^{32}$P]dATP and 1.5 μl of Klenow (1 U/μl in KDB).

4  Immediately aliquot 3.5 μl of each annealing reaction to the corresponding tube sets labeled A1–T1, etc.

5  To each A1–A$n$, add 3 μl of A Klenow termination mix; to each C1–C$n$, add 3 μl of C Klenow termination mix, etc.

6  Tap the tube to mix and incubate at 50°C for 10 min.

7  Add 4 μl of dye/formamide/EDTA solution. Vortex and incubate uncapped at 100°C for 10 min.

8  Set up, load and run the polyacrylamide gel as described in *Protocol 25*.

*Protocol 23.  Klenow-catalyzed radiolabeled DNA sequencing*

# Protocol 24. **Modified T7-catalyzed radiolabeled DNA sequencing**

## Reagents

0.1 M Dithiothreitol (DTT), store at −20°C
Dye/formamide/EDTA solution
Modified T7 DNA polymerase 10:01 TE buffer
mT7 DNA polymerase (Sequenase$^{TM}$)
5× T7 Labeling mix
[α-$^{32}$P]dATP TBq/mmol 800 Ci/mmol⚠

10× T7 Sequencing buffer
T7 Termination mixes

## Equipment

Microcentrifuge
Water bath

## Procedure

1 Set up an annealing reaction for each clone to be sequenced:
   - 2.0 μl of distilled water
   - 1.5 μl of 10× T7 Sequencing buffer
   - 1.5 μl of primer (2.5 ng/μl)
   - 5.0 μl of DNA (0.2 μg/μl)
   to give a final volume of 10.0 μl.

2 Incubate the annealing reactions at 55°C for 5 min, and then at 37°C for 30 min. During the annealing incubations, prepare a 7 M urea, 6% poly-acrylamide gel (38× 50 cm) as described in *Protocol 25*.

3 To each annealing reaction, add 1.0 μl of 0.1 M DTT, 2.0 μl of 5× T7 labeling mix, 1.5 μl of [α-$^{32}$P]dATP (800 Ci/mmol), 2.0 μl of mT7 DNA

## Notes

This procedure takes about 1 hour.

polymerase [1:8 in TE (10:0.1) buffer] and incubate at room temperature for 5 min.

4  Set up four 0.5 ml microcentrifuge tubes per annealing reaction labeled as follows: A1, C1, G1, T1, etc. To each A1–T1, etc., add 3.5 µl of the corresponding annealing/labeling reaction.

5  To each A1–A*n*, add 2.5 µl of A T7 termination mix; to each C1–C*n*, add 2.5 µl of C T7 termination mix, etc.

6  Tap the tube to mix and incubate at 37°C for 5 min.

7  Add 4 µl of dye/formamide/EDTA solution. Vortex and incubate uncapped at 100°C for 10 min. Set up, load and run the polyacrylamide gel as described in *Protocol 25*.

*Protocol 24. Modified T7-catalyzed radiolabeled DNA sequencing*

## Reagents

40% Acrylamide/bisacrylamide (40% A+B)▽

15% Ammonium persulfate (APS)

10× MTBE (modified Tris-borate-EDTA buffer)

Silanizing reagent▽

TEMED (*N,N,N',N'*-tetramethylethylenediamine) store protected from light at 4°C (Kodak T-7024)▽

Urea (Gibco/BRL)

## Equipment

Automatic pipettor with flat gel loading tips

Cassette

Fan

Flat gel loading tips

Glass plates with 0.15 mm spacers

Hoefer gel dryer

Kodak XRP-1 film

Kodak GBX developer

Plastic wrap

40 cm×20 cm Sheet of 3MM Whatman paper

Vertical electrophoresis apparatus

Well-forming comb

## Procedure

1  Prepare 8 M urea, polyacrylamide gels according to the following recipe (100 ml), depending on the desired percentage:

## Notes

This procedure takes about 4 hours.

①  Prior to taping, the notched, front glass plate should be treated with a small amount of silanizing reagent and then rinsed with double-distilled water.

|                        | 4%      | 4.75%   | 5%       | 6%      |
|------------------------|---------|---------|----------|---------|
| Urea                   | 48 g    | 48 g    | 48 g     | 48 g    |
| 40% A + B              | 10 ml   | 11.9 ml | 12.5 ml  | 15 ml   |
| 10× MTBE               | 10 ml   | 10 ml   | 10 ml    | 10 ml   |
| Double-distilled water | 42 ml   | 40.9 ml | 39.5 ml  | 37 ml   |
| 15% APS                | 500 µl  | 500 µl  | 500 µl   | 500 µl  |
| TEMED                  | 50 µl   | 50µl    | 50 µl    | 50 µl   |

2 Combine the urea, MTBE buffer and water and incubate for 5 min at 55°C and then stir to dissolve the urea.

3 Cool briefly, add the A + B, mix and degas under vacuum for 5 min.

4 While stirring, add the APS and TEMED polymerization agents and then immediately pour in between two taped glass plates with 0.15 mm spacers.

5 Insert the well-forming comb, clamp and allow the gel to polymerize for at least 30 min.

6 Prior to loading, remove the tape around the bottom of the gel and the well-forming comb. Assemble the vertical electrophoresis apparatus by clamping the upper and lower buffer chambers to the gel plates and add 1× MTBE electrophoresis buffer to the chambers.

7 Flush the sample wells with a syringe containing 1× MTBE and, immediately prior to loading each sample, flush the well with 1× MTBE using flat gel loading tips.

9 Load 1–2 µl of sample into each well using an automatic pipettor with flat gel loading tips, and then electrophorese according the following guidelines (during electrophoresis, cool the gel with a fan):

| Termination reaction | Polyacrylamide gel | Electrophoresis conditions |
|---|---|---|
| Short | 5%, 0.15 mm × 50 cm × 20 cm | 2.25 h,  22 mA |
| Long | 4%, 0.15 mm × 70 cm × 20 cm | 8–9 h,  15 mA |
| Long | 4%, 0.15 mm × 70 cm × 20 cm | 20–24 h, 15 mA |

10 After electrophoresis, remove the buffer wells and the tape, and ease the gel plates apart. The gel should adhere to the back plate. Blot the gel to a 40 cm × 20 cm sheet of 3MM Whatman paper, cover with plastic wrap, and dry on a Hoefer gel dryer for 25 min at 80°C.

11 Place the dried gel in a cassette and expose to Kodak XRP-1 film.

12 Develop the film for 1–5 min in Kodak GBX developer, rinse in distilled water for 30 sec, fix in Kodak GBX fixer for 5 min and then rinse again in distilled water for 30 sec. Allow the film to air dry.

*Protocol 25.   Radiolabeled sequencing gel preparation, loading and electrophoresis*

## Reagents

Ampli*Taq* polymerase
Diluted *Taq* polymerase (see recipe below)
95% Ethanol: 95 ml 100% ethanol, add 5 ml of distilled water△
Fluorescent end-labeled primers; prepare a 100× primer stock
  solution (40 μM) of fluorescent end-labeled primers①②
1.0 M $MgCl_2$
5× *Taq* dilution buffer
5× *Taq* reaction buffer

1.0 M Tris–HCl, pH 9.0
5× *Taq* cycle sequencing mixes (see Appendix A: reagents
  p. 146)

## Equipment

PE Cetus Thermocycler 9600
Table-top centrifuge
0.2 ml Thin-walled reaction tubes and caps (Robbins Scientific)

## Procedure

1  Pipette 1 or 2 μl of each DNA sample (100 ng/μl for M13 templates and
   200 ng/μl for pUC templates) into the bottom of the 0.2 ml thin-walled
   reaction tubes. Use the 1 μl sample for A and C reactions, and the 2 μl
   sample for G and T reactions. Meanwhile, pre-heat the PE Cetus
   Thermocycler 9600 to 95°C.

2  Prepare the *Taq* polymerase dilution.
   • 30 μl of Ampli*Taq* (5 U/μl)
   • 30 μl of 5× *Taq* dilution buffer

## Notes

This procedure takes about 3 hours.

①  An example calculation for a dry tube of an 18-mer with
   an OD of 1.00 is shown below.
   $1.00\ OD(37\ \mu g/OD)(mol \times mer/320\ g)(10^{12}\ pmol/mol)$
   $(g/10^6\ \mu g)\ (1/18\text{-}mer)(\mu l/40\ pmol) = x\ \mu l$
   In this example x = 160 μl, and therefore 160 μl of double-
   distilled water should be added to the dried tube of fluorescent primer for a concentration of 40 μM (40 pmol/μl).
②  The current primers work optimally at the effective

- 130 μl of double-distilled water

giving 190 μl of diluted *Taq* for 24 clones.

3 Prepare the A, C, G and T base-specific mixes by adding base-specific primer and diluted *Taq* to each of the base-specific nucleotide/buffer pre-mixes. For A,C/G,T use: 60/120 μl of 5× *Taq* cycle sequencing mix; 30/60 μl of diluted *Taq* polymerase; and 30/60 μl of the respective fluorescent end-labeled primer, giving a final volume of 120/240 μl.

4 Seal the reaction tubes carefully with the strip caps, and centrifuge briefly at 1000 *g* (2500 r.p.m.). Place the tube/retainer set in the 9600 Cycler, abort the soak file program, and run the cycling program. This program will cycle the sequencing reactions for 30 cycles of seven temperatures (30 cycles of 95°C denaturation for 4 sec; 55°C annealing for 10 sec; 72°C extension for 1 min; 95°C denaturation for 4 sec; 72°C extension for 1 min; 95°C denaturation for 4 sec; and 72°C extension for 1 min), and then will link to a 4°C soak file until that program is aborted.[1]

5 Briefly centrifuge at 1000 *g* to reclaim condensation. Pool the four base-specific reactions into 250 μl of 95% ethanol.

6 Precipitate the sequencing reactions, and store the dried samples at −20°C.

concentration of 0.4 μM, however with each new fluorescent primer preparation, the optimal concentration must be determined.

## Pause points

[1] The reactions may be frozen at −20°C after cycling, prior to the pooling step.

*Protocol 26. Taq polymerase-catalyzed cycle sequencing*

# Protocol 27. Fluorescent terminator reactions

## Reagents

9.5 µl of ABI supplied pre-mix (ABI)
0.8 µM Primer
Sephadex G-50 medium, bead diameter: 50–150 µl (Pharmacia) in
   sterile distilled water

## Equipment

Centri-Sep columns (Princeton Separations)
1.5 ml Eppendorf tubes
Microcentrifuge
Microtiter plate with filter (Millipore)
Savant Speed-Vac
Table-top centrifuge
0.2 ml Thin-walled cycler tubes tubes and caps (Robbins
   Scientific)
V-bottomed microtiter plate (Dynatech)

## Procedure

1. Place 0.5 µg of single-stranded or 1 µg of double-stranded DNA in 0.2 ml thin-walled cycler tubes.

2. Add 1 µl (for single-stranded templates) or 4 µl (for double-stranded templates) of 0.8 µM primer and 9.5 µl of ABI supplied pre-mix to each tube, and bring the final volume to 20 µl with double-distilled water.

3. Centrifuge briefly and cycle as usual using the terminator program as described by the manufacturer (i.e. pre-heat at 96°C followed by 25 cycles of 96°C for 15 sec, 50°C for 1 sec, 60°C for 4 min, and then link to a 4°C hold).

## Notes

This procedure takes about 3 hours.

1. If flow does not begin immediately, apply gentle pressure to the column with a pipette bulb.

2. This protocol was developed at the *C. elegans* Genome Sequencing Center at Washington University, St. Louis, Missouri, and was conveyed to us by Dr Richard Wilson, and modified accordingly.

4  Proceed with the spin column purification using either the Centri-Sep columns or G-50 microtiter plate procedures given below.

### *Terminator reaction clean-up via Centri-Sep columns*

1  Gently tap the column to cause the gel material to settle to the bottom of the column.

2  Remove the column stopper and add 0.75 ml of distilled water.

3  Stopper the column and invert it several times to mix. Allow the gel to hydrate for at least 30 min at room temperature. Columns can be stored for a few days at 4°C; longer storage in water is not recommended. Allow columns that have been stored at 4°C to warm to room temperature before use. Remove any air bubbles by inverting the column and allowing the gel to settle. Remove the upper-end cap first and then remove the lower-end cap. Allow the column to drain completely, by gravity.①

4  Insert the column into the wash tube provided.

5  Spin in a variable-speed microcentrifuge at 1300 *g* for 2 min to remove the interstitial fluid.

6  Remove the column from the wash tube and insert it into a 1.5 ml Eppendorf tube.

7  If the reactions were performed in a cycler tube with a mineral oil overlay, carefully remove the reaction mixture (20 µl) from beneath the oil and load it on top of the gel material. Avoid picking up oil with the sample; small amounts of oil (<1 µl) in the sample will not affect results.

**103**

*Protocol 27.  Fluorescent terminator reactions*

Oil at the end of the pipette tip containing the sample can be removed by touching the tip carefully on a clean surface (e.g. the reaction tube). Alternatively, if the cycle sequencing reactions were performed in a cycler tube not requiring a mineral oil overlay, carefully load the entire reaction mixture to the top of the gel material. Use each column only once.

8  If the columns are spun in a variable-speed microcentrifuge with a fixed-angle rotor, place the column in the same orientation as it was in for the first spin – this is important because the surface of the gel will be at an angle in the column after the spin. Alternatively, perform this centrifugation in a swinging-bucket style centrifuge.

9  Dry the sample in a vacuum centrifuge. Do not apply heat. Do not overdry. If desired, reactions can be ethanol precipitated.

### *Terminator reaction clean-up via Sephadex G-50-filled microtiter format filter plates* ②

1  Sephadex settles out; therefore, you must resuspend before adding to the plate and also after approximately eight columns.

2  Add 400 µl of mixed Sephadex G-50 to each well of a microtiter filter plate.

3  Place the microtiter filter plate on top of a microtiter plate to collect water, and tape the sides so they do not fly apart during centrifugation.

4  Spin at 800 *g* for 2 min.

5  Discard water that has been collected in the microtiter plate.

6  Add 200 μl of the Sephadex G-50 to fill the microtiter plate wells.

7  Place the microtiter filter plate on top of a microtiter plate to collect water, and tape the sides so they do not fly apart during centrifugation.

8  Spin at 800 *g* for 2 min.

9  Discard water that has been collected in the microtiter plate.

10 Place the microtiter filter plate on top of a clean V-bottomed microtiter plate to collect samples, and tape the sides so they do not fly apart during centrifugation.

11 Add 20 μl of terminator reaction to each Sephadex G-50-containing well.

12 Spin at 800 g for 2 min.

13 Dry the collected effluent in a Speed-Vac for approximately 1–2 h.

*Protocol 27. Fluorescent terminator reactions*

# Protocol 28. Sequenase™-catalyzed sequencing with dye-labeled terminators

## Reagents

ABI terminator mix (Perkin-Elmer)
8.0 M Ammonium acetate
9.5 M Ammonium acetate
Diluted Sequenase™ (3.25 U/μl): undiluted Sequenase™ (United States Biochemicals) is 13 U/μl and should be diluted 1:4 with USB dilution buffer prior to use, resulting in a working dilution of 3.25 U/μl
70% Ethanol△
95% Ethanol△
10× Mn²⁺/isocitrate buffer
1.0 M MgCl₂
1.0 M MOPS, pH 7.5
2.7 M MOPS (acid form)
MOPS-acid buffer

10× MOPS buffer
3 M NaCl
10 M NaOH
2 mM α-S-dNTPs: 100 μl of 20 mM α-S-dATP, 100 μl of 20 mM α-S-dCTP, 100 μl of 20 mM α-S-dGTP, 100 μl of 20 mM α-S-dTTP, 100 μl of TE (50:1) buffer, 500 μl of double-distilled water.
TE (50:1) buffer

## Equipment

Cotton swabs or tissues
Ice-water bath
Microcentrifuge
1.5 ml Microcentrifuge tubes
Savant Speed-Vac
Vortex mixer

## Notes

This procedure takes about 2 hours.

## Procedure

### *For single-stranded reactions*

1  Add the following to a 1.5 ml microcentrifuge tube:
  • 4 μl of ss DNA (2 μg)
  • 4 μl of 0.8 μM primer

- 2 μl of 10x MOPS buffer
- 2 μl of 10x $Mn^{2+}$/isocitrate buffer

to give a final volume of 12 μl.

2  To denature the DNA and anneal the primer, incubate the reaction at 65–70°C for 5 min. Allow the reaction to cool at room temperature for 15 min, and then briefly centrifuge to reclaim condensation.

3  To each reaction, add the following reagents and incubate for 10 min at 37°C. (For more than one reaction, a pot of the reagents should be made.)
- 7 μl of ABI terminator mix
- 2 μl of diluted Sequenase™
- 1 μl of 2 mM $\alpha$-$^{35}$S dNTPs
- Double-distilled water to a final voume of 22 μl.

4  Add 25 μl of 9.5 M ammonium acetate and 150 μl of 95% ethanol to stop the reaction, and vortex.

5  Precipitate the DNA in an ice-water bath for 10 min. Centrifuge for 15 min at 12 000 $g$ in a microcentrifuge at 4°C. Carefully decant the supernatant, and rinse the pellet by adding 300 μl of 70% ethanol. Vortex and centrifuge again for 15 min, and carefully decant the supernatant.

6  Repeat the rinse step to ensure efficient removal of the unincorporated terminators. Alternatively, after the first rinse step, droplets of supernatant can be removed by carefully absorbing them with a cotton swab or a rolled up tissue.

*Protocol 28. Sequenase™-catalyzed sequencing with dye-labeled terminators*

7 Dry the DNA for 5–10 min (or until dry) in the Savant Speed-Vac, and store the dried reactions at −20°C.

## *For double-stranded reactions*

1 Add the following to a 1.5 ml microcentrifuge tube:
- 5 µl of dsDNA (5 µg)
- 4 µl of 1.0 M NaOH
- 3 µl of double-distilled water

to give a final volume of 12 µl.

2 Incubate the reaction at 65–70°C for 5 min, and then briefly centrifuge to reclaim condensation.

3 Add the following reagents to each reaction, vortex and briefly centrifuge:
- 3 µl of 8 µM primer
- 9 µl of double-distilled water
- 4 µl of MOPS-acid buffer

to give a final volume of 28 µl.

4 To each reaction, add the following reagents and incubate for 10 min at 37°C. (For more than one reaction, a pot of the reagents should be made.)
- 4 µl of 10x $Mn^{2+}$/isocitrate buffer
- 6 µl of ABI terminator mix
- 2 µl of diluted Sequenase$^{TM}$
- 1 µl of 2 mM $\alpha$-S dNTPs
- Double-distilled water to give a final volume of 47 µl.

5 Add 60 µl of 8 M ammonium acetate and 300 µl of 95% ethanol to stop the reaction, and vortex.

6  Put the tubes in an ice-water bath for 10 min to complete DNA precipitation. Centrifuge for 15 min at 12 000 $g$ in a microcentrifuge at 4°C. Carefully decant the supernatant, and rinse the pellet by adding 300 μl of 70% ethanol. Vortex and centrifuge again for 15 min, and carefully decant the supernatant.

7  Repeat the rinse step to ensure efficient removal of the unincorporated terminators. Alternatively, after the first rinse step, droplets of supernatant can be removed by carefully absorbing them with a cotton swab or a rolled up tissue.

8  Dry the DNA for 5–10 min (or until dry) in the Savant Speed-Vac.

*Protocol 28. Sequenase™-catalyzed sequencing with dye-labeled terminators*

# Protocol 29. **Fluorescent DNA sequencing**

## Reagents

10× ABI TBE
1% Bind-silane (LKB)
0.5 M EDTA, pH 8.0
100 mM EDTA, pH 8.0 saturated with Blue dextran
100% Ethanol⚠
FE (formamide/EDTA)⚠

## Equipment

ABI 373A DNA Sequencer
Flat-tipped gel-loading pipette tips (Sorenson)
Table-top centrifuge
8 M Urea, 4.75% polyacrylamide gels
Vortex mixer
36-Well-forming comb

## Procedure

1 Prepare 8 M urea, 4.75% polyacrylamide gels, as described in *Protocol 25*, using a 36-well-forming comb. Alternatively, the recipe can be scaled up to 1 l.

2 Prior to loading, remove the tape from around the entire gel and carefully clean the outer surface of the gel plates with hot water. Rinse the glass with distilled water and then with ethanol, and allow the ethanol to evaporate.

3 Assemble the gel plates into an ABI 373A DNA Sequencer by placing

## Notes

This procedure takes about 2 hours.

①  This Macintosh program and the related files are available from http://dna1.chem.uoknor.edu/ as a Stuffit 1.5.1, binhexed file.

the plates on the ledge in the bottom buffer well and clamping the gel into place with the black clamps attached to the laser stop.

4  Check the glass plates by closing the ABI lid and selecting 'Start Pre-run' and then 'Plate Check' from the ABI display. Adjust the PMT on the ABI display ('Calibration', 'PMT') so that the lower scan (usually the blue) line corresponds to an intensity value of 800–1000 as displayed on the Macintosh computer data collection window. If the baseline of four-color scan lines is not flat, reclean the glass plates.

5  Attach the top buffer chamber and the alignment brace, and fill both buffer wells with 1× ABI TBE electrophoresis buffer (dilute from 10× stock). Affix the aluminum heat distribution plate by setting it on the laser stop against the glass plates.

6  Pre-electrophorese the gel for 10–30 min by choosing 'Start Pre-run' and 'Pre-run Gel'.

7  Create a sample sheet from within the ABI data collection software by entering the names and the fluorescent mobility file ('DyePrimer {M13RP1}' for fluorescent-labeled M13 universal reverse primer, 'DyeTerm {any primer}' for AmpliTaq Terminators and 'DyeTerm{T7}-SetB' for Sequenase™ fluorescent-labeled dye terminators) to use for analysis.①

8  Prepare the samples for loading. Add 3 µl of FE to the bottom of each tube, vortex, heat at 90°C for 3 min, and centrifuge to reclaim condensation.

9  Abort the pre-electrophoresis, and flush the sample wells with

**111**

*Protocol 29. Fluorescent DNA sequencing*

electrophoresis buffer with a syringe. Using flat-tipped gel-loading pipette tips, load each odd-numbered sample. Pre-electrophorese the gel for at least 5 min, flush the wells again and then load each even-numbered sample.

10  Begin the electrophoresis (35 W for 10 h) run by selecting 'Start Run' on the ABI display and by choosing 'Begin Data Collection' from the controller box within the ABI data collection software on the Macintosh.

11  After data collection, the ABI software will automatically open the data analysis software, which will create the imaged gel file, extract the data for each sample lane, and process the data. Check the imaged gel file for sample tracking, and then transfer the results folder containing the sequence trace files to a SPARCstation 2 where the hard disk is mounted on the ether-netted Macintosh computer via NFS Share.

## Reagents

8.0 M Ammonium acetate
$[\alpha\text{-}^{32}P]dATP$ (> 3.7 Tbq/mmol, >1000 Ci/mmol)⚠
0.5 M EDTA, pH 8.0
95% Ethanol⚠
70% Ethanol⚠
1.0 M $MgCl_2$
2.7 M MOPS (acid form)
MOPS-acid buffer
0.2 M NaOH⚠

Poly(dT)-anchored primers
T7 DNA polymerase (United States Biochemicals)

## Equipment

ABI Oligonucleotide Purification Cartridges (Perkin Elmer)
Applied Biosystems (ABI) 391 DNA Synthesizer
Microcentrifuge
Savant Speed-Vac

## Procedure

1 Synthesize anchored poly(dT)$_{17}$ with anchors of (dA), (dC) or (dG) at the 3′ end on a DNA synthesizer and use after purification on Oligonucleotide Purification Cartridges.

2 For sequencing with anchored primers, denature 5–10 µg of plasmid DNA in a total volume of 50 µl containing 0.2 M NaOH and 0.16 mM EDTA, pH 8.0 by incubation at 65°C for 10 min.

3 Add the three poly(dT)-anchored primers (2 pmol of each) and

## Notes

This procedure takes about 2 hours.

immediately place the mixture on ice. Neutralize the solution by adding 5 μl of 5 M ammonium acetate pH 7.0.

4  Precipitate the DNA by adding 150 μl of cold 95% ethanol and wash the pellet twice with cold 70% ethanol.

5  Dry the pellet for 5 min and then resuspend in MOPS-acid buffer.

6  Anneal the primers by heating the solution for 2 min at 65°C followed by slow cooling to room temperature for 15–30 min.

7  Perform sequencing reactions, using modified T7 DNA polymerase and [α-$^{32}$P]dATP (> 1000 Ci/mmol) using the protocol described in *Protocol 24*.

*Protocol 30.  Double-stranded sequencing of cDNA clones*

# Protocol 31. cDNA sequencing based on PCR and random shotgun subcloning

## Reagents

AmpliTaq DNA polymerase (Perkin Elmer)
100 pmol each primer
Glycerol
Isopropyl alcohol/dry-ice or saturated aqueous NaCl/dry-ice bath
1.0 M KCl
Klenow DNA polymerase (New England Biolabs)
Low melting agarose gel
1.0 M MgCl$_2$
Nucleotides: dNTP set (Pharmacia)

T4 DNA polymerase (New England Biolabs)
T4 Polynucleotide kinase (United States Biochemicals)
1.0 M Tris–HCl, pH 8.5

## Equipment

Agarose gel electrophoresis apparatus
Nebulizer (IPI Medical Products)
PE-Cetus 48 tube DNA Thermal Cycler
0.5 ml Snap-cap tubes

## Procedure

1 Incubate four 100 μl PCR reactions, each containing approximately 100 ng of plasmid DNA, 100 pmol of each primer, 50 mM KCl (dilute from 1 M stock), 10 mM Tris–HCl pH 8.5 (dilute from 1 M stock), 1.5 mM MgCl$_2$ (dilute from 1 M stock), 0.2 mM of each dNTP (dilute from 100 mM stock), and 5 units of PE-Cetus AmpliTaq in 0.5 ml snap-cap tubes for 25 cycles of 95°C for 1 min, 55°C for 1 min and 72°C for 2 min in a PE-Cetus 48 tube DNA Thermal Cycler.

2 After pooling the four reactions to obtain sufficient quantities of PCR

## Notes

This procedure takes about 24 hours depending on the time for the ligation and gel extraction.

product for the subsequent steps, the aqueous solution containing the PCR product is placed in a nebulizer, brought to 2.0 ml by adding approximately 0.5–1.0 ml of glycerol, and equilibrated at –20°C by placing it in either an isopropyl alcohol/dry-ice or saturated aqueous NaCl/dry-ice bath for 10 min.

3 The sample is nebulized at –20°C by applying 25–30 p.s.i. (1.8–2.0 bar) nitrogen pressure for 2.5 min. Following ethanol precipitation to concentrate the sheared PCR product, the fragments were blunt-ended and phosphorylated by incubation with the Klenow fragment of *E. coli* DNA polymerase, T4 DNA polymerase and T4 polynucleotide kinase as described in *Protocol 11*. Fragments in the 0.4–0.7 kb range were obtained by elution from a low melting agarose gel.

*Protocol 31. cDNA sequencing based on PCR and random shotgun subcloning*

# V  ADDITIONAL METHODS

## Methods available

### Polymerase chain reaction (PCR) (see *Protocol 32*)

The amplification of DNA fragments using the polymerase chain reaction [1] is performed in either the Perkin-Elmer Cetus DNA Thermal Cycler or the Perkin-Elmer Cetus Cycler 9600, by adding the following reagents to either a 0.5 ml thin-walled tube or a 0.2 ml tube, respectively: a small amount of the template DNA (typically cosmid, plasmid or genomic DNA), the two primers flanking the region to be amplified, nucleotides, buffer and Ampli*Taq* DNA polymerase. The cycling protocol consists of 25–30 cycles of three temperatures: strand denaturation at 95°C, primer annealing at 55°C, and primer extension at 72°C, typically 30 sec, 30 sec and 60 sec for the DNA Thermal Cycler and 4 sec, 10 sec and 60 sec for the Thermal Cycler 9600. For reactions performed in the DNA Thermal Cycler, the reaction mixtures are overlaid with two drops of mineral oil prior to temperature cycling to eliminate liquid evaporation and condensation. This is not necessary for the Thermal Cycler 9600, which is equipped with a heated lid, maintained at 100°C, that closely contacts the sample tube caps and eliminates liquid evaporation and condensation. After PCR, aliquots of the mixture typically are loaded on to an agarose gel and electrophoresed to detect the ampli-

## References

1. Mullis, K.B. and Faloona, F.A. (1987) *Methods Enzymol.* **155**:335.
2. Starr, L. and Quaranta, V. (1992) *BioTechniques*, **13**:612.
3. Sambrook, J., Fritsch, E.F. and Maniatis, T. (1989) *Molecular Cloning: A Laboratory Manual*. Cold Spring Harbor Laboratory Press, New York.
4. Tahara, T., Kraus, J.P. and Rosenberg, L.E. (1990) *BioTechniques*, **8**:366.
5. Kusukawa, N., Uemori, T., Asada, K. and Kato, I. (1990) *BioTechniques*, **9**:66.
6. Bodenteich, A., Chissoe, S., Wang, Y.F. and Roe, B.A. (1994) In *Automated DNA Sequencing and Analysis Techniques* (C. Venter, ed.), p. 42, Academic Press, London.
7. Beckman Oligo 1000 Synthesizer manual.
8. ABI 392 Synthesizer manual.
9. Beaucage, S.L. and Caruthers, M. (1981) *Tetrahedron Lett.*, **22**:1859.
10. Matteucci, M.D. and Caruthers, M.H. (1981) *J. Am. Chem. Soc.* **103**:3185.
11. Adams, S.P., Kavka, K.S., Wykes, E.J., Holder, S.B. and Galluppi, G.R. (1983) *J. Am. Chem. Soc.* **105**:661.
12. Eadie, J.S. and Davidson, D.S. (1987) *Nucleic Acids Res.* **15**:8333.

fied product. In some instances, where the yield from a single PCR is insufficient, the reaction is ethanol precipitated, resuspended and an aliquot is used as template for a second round of PCR amplification.

### *Purification of PCR fragments for cloning* (see *Protocol 33*)

After an aliquot of the PCR mixture is analyzed on an agarose gel, the remainder of the reaction is concentrated by ethanol precipitation, resuspended in buffer and subjected to a simultaneous fill-in/kinase reaction with the Klenow fragment of *E. coli* DNA polymerase and T4 polynucleotide kinase, the four deoxynucleotides and rATP [2]. The reaction then is loaded on to a preparative 1, 1.5 or 2% low-melting temperature agarose gel, depending on the size(s) of the fragment(s) as determined above and, after minimal separation is achieved between the product(s) and the excess primers, the DNA fragments are excised and eluted. After concentration by ethanol precipitation, amplified DNA fragments are ligated into blunt-ended cloning vector, such as *Sma*I-linearized, dephosphorylated double-stranded M13 replicative form (RF) or pUC.

It should be noted that several other methods to purify the DNA fragments for cloning have been investigated. These include standard ethanol/acetate precipitation [3], a 50% ethanol precipitation (R.K. Wilson, personal communication), spin-column purification [4] and precipitation with polyethylene glycol (PEG) [5]. The first three methods did not remove sufficient unincorporated primer and, during

13. Farrance, I.K., Eadie, J.S. and Ivarie, R. (1989) *Nucleic Acids Res.* **17**:1231.
14. Smith, L.M., Kaiser, R.J., Sanders, J.Z. and Hood, L.E. (1987) *Methods Enzymol.* **155**:260
15. Sawadogo, M. and Van Dyke, M.W. (1991) *Nucleic Acids Res.* **19**:674.

## Protocols provided

**32.** *Polymerase chain reaction (PCR)*
**33.** *Purification of PCR fragments for cloning*
**34.** *Preparation of Sma*I-linearized, dephosphorylated double-stranded M13 RF cloning vector*
**35.** *Synthesis and purification of oligonucleotides*
**36.** *Rapid hybridization of complementary M13 inserts*

the subsequent ligation of the DNA fragment, the primers apparently competed for the blunt-ended vector during ligation because the efficiency of ligation was significantly lower and the vast majority of recombinant clones contained only primer-derived inserts. Precipitation by PEG resulted in only an extremely small DNA pellet, removing the PEG supernatant is difficult and yields of PCR product were variable.

### Preparation of SmaI-linearized, dephosphorylated double-stranded M13 RF cloning vector (see *Protocol 34*)

In this method, adapted from ref. 6, several small batches of M13 RF double-stranded DNA are mixed with buffer, *Sma*I restriction endonuclease and calf intestinal alkaline phosphatase (CIAP) and incubated for 2–4 h at 37°C. The reactions then are pooled, ethanol precipitated and resuspended in buffer to yield a final concentration of about 10 ng/μl. After characterization to determine the optimal concentration for shotgun cloning ligations and to assess the efficiency of *Sma*I digestion and CIAP dephosphorylation, aliquots of linearized vector are stored frozen at −70°C.

### Synthesis and purification of oligonucleotides (see *Protocol 35*)

Oligonucleotide primers, either to close gaps in a cosmid or plasmid sequencing project or for PCR, are chosen by manual observation using the following criteria:

(i) a relatively even base distribution (~50% GC) is desired, with no obvious repeated motifs;

(ii) when possible, the 3′ end of the primer should contain either a G or a C residue;

(iii) to determine if the sequence chosen is unique, the sequence is compared with the available cosmid or plasmid sequence using Findpatterns.

Alternatively, several computer programs, the SPARCstation-based ospX, the Macintosh-based HYPER PCR and Primer or the VAX-based Primer programs, could be employed and yield similar results.

Oligonucleotides are synthesized according to manufacturer's procedures on either the Beckman Oligo 1000 [7] or the ABI 392 synthesizer [8] using the phosphoramidite chemistry [9]. The desired oligonucleotide sequence is entered into the respective synthesizer, the reagent bottles are attached, and the column is inserted which contains the respective 3′ nucleotide base-specifically linked by the 3′-OH group to the solid support, controlled-pore glass (CPG) silica beads [10, 11]. The 5′-OH group of the base is blocked with a dimethyloxytrityl (DMT) group. Typically, the columns are purchased for 30 nmol of primer. For large-scale synthesis, for laboratory stocks of universal primers, columns for 1 µmol synthesis are obtained, and can only be used on the ABI 392. The synthesizer automatically performs a cycle of base addition which consists of

5′-detritylation to remove the 5′-DMT blocking group with trichloroacetic acid and dimethylchloride, activating the phosphoramidite nucleoside monomer with tetrazole and coupling the activated phosphoramidite nucleoside to the column, capping chains that were not coupled during the previous step by acylation of the 5′-OH end with acetic anhydride and 1-methylimidazole [12, 13] and oxidizing the internucleotide phosphate linkage from the phosphite ester to the more stable phosphotriester with iodine and water [8]. After each step in the synthesis, the column is washed with acetonitrile. For the synthesis of fluorescent 5′-end-labeled oligonucleotides, the last base added has an aminolink on the 5′ end [14]. After synthesis, the oligonucleotide is removed from the solid support, the protecting groups are removed, and the primer is used directly after concentration by butanol precipitation [15].

### Rapid hybridization of complementary M13 inserts

(see *Protocol 36*)

M13 clones carrying complementary inserts (i.e. in the opposite orientation) may be screened rapidly by clone-to-clone hybridization, followed by analysis on an agarose gel.

# Protocol 32. **Polymerase chain reaction (PCR)**

## Reagents

1–2% Agarose gel

Ampli*Taq* DNA polymerase (Perkin Elmer)

2 mM dNTPs: 100 µl of 20 mM dATP, 100 µl of 20 mM dCTP, 100 µl of 20 mM dGTP, 100 µl of 20 mM dTTP, 100 µl of TE (50:1) buffer, 500 µl of double-distilled water

0.5 M EDTA, pH 8.0

100 mM EDTA, pH 8.0: 20 ml of 0.5 M EDTA, pH 8.0, 80 ml of double-distilled water

1.0 M KCl

1.0 M $MgCl_2$

Mineral oil

10 M NaOH

10× PCR buffer

Primers 40 µM

TE (50:1) buffer

1.0 M Tris–HCl, pH 7.6

1.0 M Tris–HCl, pH 8.5

## Equipment

Agarose gel electrophoresis apparatus

0.5 ml Flat-topped microcentrifuge tube

Perkin-Elmer Cetus DNA Thermal Cycler

## Procedure

1 Add the following reagents to a 0.5 ml flat-topped microcentrifuge tube:
  - 1 µl of target DNA (10–20 ng)
  - 2.5 µl of each primer (40 µM )
  - 1 µl of Ampli*Taq* DNA polymerase (5 U)
  - 10 µl of 2 mM dNTPs (2 mM each dNTP)
  - 10 µl of 10× PCR buffer
  - 75.5 µl of double-distilled water
  to give a total volume of 100 µl.

## Notes

This procedure takes about 4 hours.

2 Cover the reaction with two drops of mineral oil, add a drop of oil to the heat-block well to ensure good contact between the heat-block and the tube, and place the tube in the wells of a Perkin-Elmer Cetus DNA Thermal Cycler which has been pre-heated to 95°C.

3 Abort the soak file program and begin the thermal-cycle program which has 25 cycles of a three-temperature program and is linked to a 4°C soak file, which will hold indefinitely: 95°C for 1 min, 55°C for 1 min, 72°C for 2 min.

4 Analyze a 10 μl aliquot on a 1–2% agarose gel.

# Protocol 33. Purification of PCR fragments for cloning

## Reagents

10× Agarose gel loading dye
Buffer saturated phenol
95% Ethanol/0.12 M sodium acetate (ethanol/acetate)⚠
100% Ethanol⚠
TE (10:0.1) buffer

## Equipment

Agarose gel elecrophoresis apparatus
Clean razor blade
1.0% Low-melting temperature agarose gel
UV light source⚠

## Procedure

1 Ethanol precipitate the PCR reaction by adding 2.5 vol. of 95% ethanol containing 0.12 M sodium acetate, pH 4.8.

2 Perform the combined fill-in kinase reactions described in *Protocol 11*.

3 Add 5 µl of agarose gel loading dye and load the reaction into a well of a 1.0% low-melting temperature agarose gel. Electrophorese for 30–60 min at 100–120 mA, and then excise the desired band visualized under UV light with a clean razor blade.

4 Elute the DNA from the gel by standard freeze–thaw methods, followed by a phenol extraction and concentrate by ethanol precipitation.

## Notes

This procedure takes about 3 hours.

5 Resuspend the dried DNA in 10 μl of TE (10:0.1) buffer. Use this DNA in a standard blunt-ended ligation reaction. Typically, use 2–3 μl of this DNA in a 10 μl ligation reaction with 20 ng of pUC 18/SmaI-CIAP, although this will depend on the yield of amplified DNA from the PCR reaction.

*Protocol 33.   Purification of PCR fragments for cloning*

## Reagents

Calf intestinal alkaline phosphatase (CIAP, Boehringer Mannheim)

0.5 M EDTA, pH 8.0

NEB buffer #4: (500 mM potassium acetate, 200 mM Tris-acetate, 100 mM magnesium acetate, 10 mM dithiothreitol, pH 7.9) included with *Sma*I from New England Biolabs

*Sma*I (16 U/µl, New England Biolabs 141L)

TE (10:0.1) buffer

1 M Tris–HCl, pH 7.6

## Equipment

10–20 Tubes

## Procedure

1 Prepare 10–20 tubes with the following:
   - 5 µg of M13 RF
   - 2 µl of NEB buffer #4
   - 4.5 µl of *Sma*I (16 U/µl)
   - 3 µl of CIAP
   - double-distilled water to 20 µl.

2 Incubate at 37°C for 2–4 h.

3 Pool the reactions, phenol extract, ethanol precipitate and resuspend the dried DNA in TE (10:0.1) buffer to a final concentration of 10 ng/µl.

## Notes

This procedure takes about 4 hours.

*Protocol 34. Preparation of* SmaI-*linearized, dephosphorylated double-stranded M13 RF cloning vector*

## Reagents

ABI reagents:
Aminolink-2, FAM-NHS, JOE-NHS, ROX-NHS,
TAMRA-NHS
10× Agarose gel loading dye
Amberlite MB resin
8.0 M Ammonium acetate
30% Ammonium hydroxide▽
$n$-Butanol
De-ionized formamide▽
Dimethylformamide▽
Dry acetonitrile
0.5 M EDTA, pH 8.0
95% Ethanol/0.12 M sodium acetate (ethanol/acetate)▽
95% Ethanol▽
Fluorescent dye: four dyes named FAM, JOE, ROX and
TAMRA come from ABI, each as a solution in 60 µl DMSO.
3 M NaCl
0.5 M $NaHCO_3$/$Na_2CO_3$ pH 9.0: adjust 0.5 M $NaHCO_3$ to pH 9.0
with 0.5 M $Na_2CO_3$
10 M NaOH▽
20× TAE buffer
TE (10:0.1) buffer
1 M Tris–HCl pH 7.6

## Equipment

ABI 392 or Beckman Oligo 1000 automated DNA synthesizers
Agarose gel electrophoresis apparatus
Cling film (polyvinylchloride film)
4× 1 µmol Columns
1 ml Cuvette
20× 1.5 ml Eppendorf tubes
Ice
Large Falcon tube
Microcentrifuge
20% Polyacrylamide gels with 7 M urea
Razor blade
1.5 ml Screw-cap microcentrifuge tubes
Sephadex G-25 columns
Siliconized microcentrifuge tubes
Sharpie
Savant Speed-Vac
Spectrophotometer
Syringe
2 dram Vial
Vortex mixer
55°C Water bath
Whatman 3MM paper

## Procedure

1 Synthesize the oligonucleotide on either the ABI 392 or the Beckman Oligo 1000 automated DNA synthesizers according to manufacturers' instructions.

2 After the cycles of base addition are complete, the ABI 392 automatically detaches the synthesized oligonucleotide from the solid support by adding ammonium hydroxide and transfers the mixture into a 2 dram vial. For the Beckman Oligo 1000, to detach the oligonucleotide from the solid support remove the column and affix it to a screw-cap tube containing 1 ml of concentrated ammonium hydroxide and, with a syringe attached to the fluted end of the column, draw the ammonium hydroxide into the column. Allow this assembly to remain at room temperature for at least 1 h, no longer than 2 h and, after about 30 min, mix the solution with the syringe. Push the liquid out of the column with the syringe into the tube.

3 Incubate the oligonucleotide in ammonium hydroxide at 70°C for at least 2 h (for 1 μmol synthesis on the ABI 392, incubate overnight) to deprotect the bases.

4 Transfer 100 μl aliquots of the mixture into nine microcentrifuge tubes, add 1.25 ml of *n*-butanol, vortex twice for 10 sec, and centrifuge at 4°C for 10 min at 13 000 *g*.

5 Decant and dry in the Savant Speed-Vac until dry (at least 2 h or overnight).

6 Add 115 μl of TE (10.0.1) buffer into the first tube, resuspend by pipetting

## Notes

This procedure takes about 20 hours overall.

The Beckman oligonucleotide 1000 DNA synthesizer can also be used to synthesize the primers at the 1 μmol scale. In this case, the Aminolink-2 reagent is again dissolved in 3.3 ml of dry acetonitrile and then quickly transferred to the Beckman X bottle and placed on the X port. On START SYNTHESIS, the relevant sequence is chosen, synthesis SET SCALE (mmol) – 1000 and FINAL DETRITYLATION – NO set. On completion, the aminolink-oligonucleotide is manually cleaved from the column using concentrated (30%) ammonium hydroxide 1 ml, the cleaving step taking 1 h to complete. The resulting solution is then transferred to a 2 dram vial, the volume brought up to 2 ml with concentrated ammonium hydroxide and then left at 70°C for 12 h. The remaining steps are as mentioned in the protocol.

*Protocol 35. Synthesis and purification of oligonucleotides*

up and down, and then transfer into the second tube, etc., until the dried oligonucleotide in all nine tubes is contained in one tube in about 110 µl.

7 Remove 10 µl of oligonucleotide and dilute with 990 µl of double-distilled water and read the $A_{260}$ in a 1 ml cuvette. The amount of oligonucleotide in the solution in the cuvette is $100 \times A_{260}$, but the amount of oligonucleotide in the remaining 100 µl of solution is $10 \times A_{260}$.

## For synthesis of fluorescent 5′-end-labeled oligonucleotides

Primers were synthesized on an ABI 392 DNA Synthesizer at the 1 µM synthesis scale with the final detritylation and the end/cleave program DMT ON, AUTO (END-CE) employed (i.e. no final detritylation step, automatic cleaving of the oligonucleotide from the column delivering a final volume of 2 ml of concentrated ammonium hydroxide). The final base to be added is the Aminolink-2 diamine used to couple the fluorescent dye to the oligonucleotide primer. The procedure employed for the synthesis and purification of the aminolink-primers is as follows.

Edit the primer sequence into the synthesizer, the final 5′ end base being the Aminolink-2, for example 5′ 5CA GGA AAC AGC TAT GAC C 3′, the 5 representing the Aminolink-2 reagent placed in the bottle 5 position.

1 Dissolve the Aminolink-2 in 3.3 ml of dry acetonitrile and place it in the bottle 5 position.

2 Place fresh concentrated (30%) ammonium hydroxide solution in the ammonium hydroxide bottle (bottle 10) if needed.

3 START SYNTHESIS on the ABI, selecting the relevant sequences to be

synthesized, DMT ON, AUTO (END CE) and EXECUTE ABI BEGIN – YES if synthesizer has not been used in the last 6 h.

4  Remember to place a 2 dram vial at the outlet port for collection of the synthesized oligonucleotide.

5  When the synthesis and cleaving steps are complete, remove the vial, which should contain 2 ml of solution, cap it tightly and leave at 70°C for 12 h to remove the base-protecting groups.

6  Precipitate the DNA using the method of Sawadogo and Van Dyke [15] in which to 1 part DNA/ammonium hydroxide solution 10 parts of n-butanol are added. This is best achieved by aliquoting the DNA solution into 20× 1.5 ml microcentrifuge tubes of about 100 µl each and adding 1.25 ml n-butanol to each tube, vortex each for 10 sec then centrifuge for 20 min at 13 000 g at 4°C.

7  Pour off the supernatant, drain the tube on a paper towel for 5 min then dry the pellet in a Savant Speed-Vac for at least 2 h (overnight is preferable to be certain that no ammonium hydroxide or n-butanol remains).

8  To pool together and desalt, the following procedure is employed.
   (i)  Combine two samples into one (therefore the 20 microcentrifuge tubes are combined, resulting in 10 tubes) by dissolving one of the two samples in 40 µl of 1 M NaCl and transferring it into the second tube. Further wash the first tube with 32 µl of 1 M NaCl and again transfer to the second tube, giving a total volume of 72 µl.
   (ii)  Add 84 µl of 95% ethanol and briefly vortex.

*Protocol 35.  Synthesis and purification of oligonucleotides*

(iii) Add a further 84 µl of 95% ethanol, vortex and then leave at −20°C for 30 min to precipitate.

(vi) Centrifuge the precipitated samples for 20 min at 13 000 $g$ and 4°C, pour off the supernatant and dry the samples in a Savant Speed-Vac for 15 min.

9 The aminolink-primer oligonucleotide is now ready for coupling with the four fluorescent dyes and may be stored at −20°C until ready to couple.

The next step in the synthesis of the labeled primers is the coupling reaction between the fluorescent dye and the aminolinked-oligonucleotide primer followed by its eventual purification. The four dyes named FAM, JOE, ROX and TAMRA come from ABI, each as a solution in 60 µl DMSO. These dyes (which once diluted must all be used) are further diluted with 440 µl of DMF to give a final volume of 500 µl (50 µl/reaction) and are therefore enough for 10 reactions each. Therefore, it is best to prepare 4 × 1 µM columns worth of aminolinked-oligonucleotide primers as described above, which then will be distributed into 40× 1.5 ml micro-centrifuge tubes, 10 reactions for each dye. The following procedure was therefore employed to prepare the fluorescent-labeled primers from the aminolinked-oligonucleotides.

1 First prepare 4× 1 µM columns worth of aminolinked-oligonucleotide as described above, yielding 40× 1.5 ml microcentrifuge tubes of oligonucleotide ready for coupling.

2 Dissolve each of the samples in 50 µl of double-distilled water and 50 µl of 0.5 M $NaHCO_3/Na_2CO_3$ pH 9.0.

3 Add to each of the four dyes 440 μl of DMF (to give a final volume of 500 μl), then briefly vortex and centrifuge.

4 From each dye, add 50 μl to each of 10 pre-dissolved aminolinked-oligonucleotide reactions and leave overnight at room temperature in the dark (covered with aluminum foil is sufficient).

5 Pool each of the 10 samples together and elute through a Sephadex G-25 column to remove any unreacted dye as follows.
   (i)   Prepare four Sephadex G-25 columns by eluting with 100 ml of 0.1 M ammonium acetate (dilute from 8.0 M stock).
   (ii)  Apply each sample to the column and elute with 0.1 M ammonium acetate, collecting the leading colored band, the second colored band being the unreacted dye.
   (iii) Aliquot the primer fractions into approximately 400 μl lots and precipitate with 1 ml of ethanol/acetate at −20°C for 2 h.
   (iv)  Centrifuge the samples for 20 min at 4°C and 13 000 g, pour off the supernatant and dry the samples in a Savant Speed-Vac.

6 To purify the dye-labeled primers from the unlabeled primers the samples were electrophoresed on a 20% polyacrylamide gel as follows.
   (i)   For each of the three primers JOE, ROX and TAMRA prepare one, and for FAM prepare two 0.3 mm/20% polyacrylamide gels with 10 wells each capable of holding a minimum of 25 μl.
   (ii)  Pool each of the four sets of primers together in 200 μl of de-ionized formamide/50 μl of double-distilled water.
   (iii) Apply 25 μl of dye solution to each of the 10 wells of the respective gels and electrophorese at 2500 V, 22–25 mA for 2.5 h.

*Protocol 35. Synthesis and purification of oligonucleotides*

(iv) Pry the gels apart and with a razor blade, cut away the colored material, place in a large Falcon tube with 2 ml of 1× TAE (dilute from 20× stock) and leave overnight at 37°C.

(v) Remove the solution and aliquot evenly into two samples of approximately 1 ml each and then wash the residue with 2 ml of 0.1 M ammonium acetate (dilute from 20× stock).

(vi) Desalt each of the two samples for each of the four dye-labeled primers by eluting with 0.1 M ammonium acetate through a Sephadex G-25 column (prepared as above), again collecting the colored band.

(vii) Pool the fractions, measure the $A_{260}$ and $A_x$ (FAM, 494 nm; JOE, 527 nm; ROX, 586 nm; TAMRA, 558 nm) (the $A_{260}$ in the 0.1 cm cuvette with the UV lamp and the $A_x$ in the 1 cm cell with the VIS lamp) and then aliquot the solutions into 1.5 ml microcentrifuge tubes, 300 µl per tube. Record the total volume to calculate the OD (see Appendix A).

(viii) Dry the samples in a Savant Speed-Vac overnight or leave in a drawer in the dark until dry and then store at –20°C.

## *PAGE purification of synthetic, fluorescent 5′-end-labeled oligonucleotides*

1 Remove DNA collection vial from DNA synthesizer following automatic cleavage from the column. Bring total volume up to 4 ml with fresh concentrated NH₄OH. Cap tightly and place at 55°C for 4–12 h.

2 Remove the vial from the 55°C water bath and place on ice for 10–15

## Notes

This procedure takes about 1–2 hours prior to the cleavage step.

① BP = 8 nt, XC = 28 nt for a 20% gel.

min. Transfer the sample to three siliconized microcentrifuge tubes and dry under vacuum for 6–10 h (until completely dry).

3   Dissolve contents of one tube in 100 ml of double-distilled water. Determine concentration of the sample by measuring absorbance at 260 nm.

4   Remove an aliquot of the sample containing approximately 2 $A_{260}$ and mix with 10–20 µl of agarose gel loading dye. This will be a sufficient amount to load four 1 cm wells (i.e. 0.5 $A_{260}$ per well).

5   Prepare a 20% polyacrylamide gel containing 7 M urea (20 cm × 40 cm × 0.4 mm). Load the samples and electrophorese at 25 mA until the slow blue dye (XC) has migrated about 14 cm from the origin (for a 17-mer).

6   Remove the top glass plate and cover the gel with Cling film. Carefully lift the Cling film and the gel off of the bottom glass plate. Flip the gel over and cover the other side with a second sheet of Cling film. Visualize the DNA bands by UV shadowing and photograph. A 17-mer will migrate halfway between the BP and XC dye bands.① Outline the oligonucleotide bands with a marker.

7   Excise the oligonucleotide bands from the gel and place each gel slice in a siliconized 0.5 ml microcentrifuge tube. Add enough TAE or water to cover the gel slice (~ 200 µl) and place in a dry 37°C incubator overnight.

8   Pool the eluant from all tubes and desalt on a small G-25 column (1–2 ml bed volume). Read the absorbance at 260 nm of all fractions. Pool the peak fractions, and re-measure the absorbance. Oligonucleotides may be used directly, or diluted for sequencing or labeling reactions, frozen in small aliquots or dried.

## Protocol 36. Rapid hybridization of complementary M13 inserts

### Reagents

0.7% Agarose gel (see *Protocol 4*)
10× Agarose gel loading dye
5× *Hin*d/DTT

### Equipment

Agarose gel electrophoresis apparatus
UV light source

### Procedure

1 Set up the hybridization reaction as follows:
  - 1.0 μl of M13 clone 1
  - 1.0 μl of M13 clone 2
  - 1.0 μl of 5× *Hin*d/DTT
  - 3.0 μl of double-distilled water
  to give a final volume of 6.0 μl. Incubate at 55°C for 30 min.①

2 Add 6 μl of 2× agarose gel loading dye (dilute from 10× stock), vortex briefly and load on 0.7% agarose gel. Electrophorese at 90 mA for 1 h.

3 Visualize DNA bands under UV light. Positives will run slower as a duplex DNA.

### Notes

This procedure takes about 2 hours.

① M13 DNA is approximately 0.5–1.0 μg/ml.

# APPENDIX A: SOLUTION RECIPES

**10× ABI TBE:** Add together 216 g of Tris base, 110 g of boric acid, 16.6 g of Na$_2$EDTA. Add water to make 2 l.

**40% Acrylamide/bisacrylamide (40% A + B):** Add together 380 g of acrylamide (Kodak 5521) and 20 g of *N,N*-methylene-bisacrylamide (Kodak 8383). Dissolve in approximately 800 ml of double-distilled water and then de-ionize by stirring with 50 g of amberlite MB-1 (Sigma MB-1A) for 1 h at room temperature. Suction filter to remove the Amberlite and adjust to a final volume of 1 l with double-distilled water. Store at 4°C.

**100 mM Adenosine triphosphate (rATP):** 619 mg of dipotassium ATP (ICN 100004), make up to 10 ml with sterile double-distilled water. Aliquot and store at −20°C.

**10× Agarose gel loading dye:** Add together 1.5 g of Ficoll (Sigma F-2637), 0.02 g of bromophenol blue (Sigma B-0126), 0.02 g of xylene cyanole FF (Kodak T-1579) and make up to 10 ml with double-distilled water. Store at −20°C.

**Alkaline lysis solution (NaOH/SDS):** Make fresh using 20 ml of 1 M NaOH (or 0.8 g), 10 ml of 10% SDS (or 1.0 g) and make up to 100 ml using double-distilled water.

**9.5 M Ammonium acetate:** 73.23 g of ammonium acetate and double-distilled water to give 100 ml.

**8.0 M Ammonium acetate:** 61.69 g of ammonium acetate and double-distilled water to make 100 ml.

**15% Ammonium persulfate (APS):** 1.5 g of APS (Kodak 11151), add double-distilled water to make 10 ml. Store at 4°C.

**Ampicillin (Amp):** 0.5 g of Amp (Sigma A-9518) and make up to 100 ml with sterile, double-distilled water. Add to media for final concentration of 100 µg/ml.

**1 mg/ml Bovine serum albumin (BSA):** 5 mg of BSA (Sigma A-9647), make up to 5 ml with sterile, double-distilled water. Aliquot and store at −20°C.

***Bst* dilution buffer:** Add together 500 µl of 1 M HEPES, pH 7.6, 100 µl of 1 M MgCl$_2$, 10 µl of 1 M DTT, 10 mg of BSA, make up to 10 ml with sterile, double-distilled water.

***Bst* 'long' termination 'A' mix:** Add together 8 μl of 0.5 mM dATP, 16.4 μl of 5 mM dCTP, 16.4 μl of 5 mM c$^7$dGTP, 16.4 μl of 5 mM dTTP, 11 μl of 5 mM ddATP, 50 μl of TE (50:1) buffer and 381.8 μl of sterile double-distilled water to give a final volume of 500 μl. Aliquot (18 μl for six reactions) and store at –70°C.

***Bst* 'long' termination 'C' mix:** Add 16.4 μl of 5 mM dATP to 8 μl of 0.5 mM dCTP, 16.4 μl of 5 mM c$^7$dGTP, 16.4 μl of 5 mM dTTP, 6.5 μl of 5 mM ddCTP, 50 μl of TE (50:1) buffer and 386.3 μl of sterile double-distilled water to give a final volume of 500 μl. Aliquot (18 μl for six reactions) and store at –70°C.

***Bst* 'long' termination 'G' mix:** Add together 16.4 μl of 5 mM dATP, 16.4 μl of 5 mM dCTP, 8 μl of 0.5 mM c$^7$dGTP, 16.4 μl of 5 mM dTTP, 7 μl of 5 mM ddGTP, 50 μl of TE (50:1) buffer and 385.8 μl of sterile double-distilled water to give a final volume of 500 μl. Aliquot (18 μl for six reactions) and store at –70°C.

***Bst* 'long' termination 'T' mix:** Add together 16.4 μl of 5 mM dATP, 16.4 μl of 5 mM dCTP, 16.4 μl of 5 mM c$^7$dGTP, 8 μl of 0.5 mM dTTP, 15 μl of 5 mM ddTTP, 50 μl of TE (50:1) buffer and 377.8 μl of sterile double-distilled water to a final volume of 500 μl. Aliquot (18 μl for six reactions) and store at –70°C.

***Bst* nucleotide extension mix:** Add 3 μl of 5 mM dCTP to 3 μl of 5 mM c$^7$dGTP, 3 μl of 5 mM dTTP, 100 μl of TE (50:1) buffer, and 891 μl of sterile double-distilled water, giving a final volume of 1 ml.

***Bst* reaction buffer:** Add together 5 ml of 1 M Tris–HCl, pH 8.5, 1.5 ml of 1 M MgCl$_2$, make up to 10 ml with sterile double-distilled water.

***Bst* 'short' termination 'A' mix:** Add 8 μl of 0.5 mM dATP to 16.4 μl of 5 mM dCTP, 16.4 μl of 5 mM c$^7$dGTP, 16.4 μl of 5 mM dTTP, 66 μl of 5 mM ddATP, 50 μl of TE (50:1) buffer and 326.8 μl of sterile double distilled water to give a final volume of 500 μl. Aliquot (18 μl for six reactions) and store at –70°C.

***Bst* 'short' termination 'C' mix:** Add together 16.4 μl of 5 mM dATP, 8 μl of 0.5 mM dCTP, 16.4 μl of 5 mM c$^7$dGTP, 16.4 μl of 5 mM dTTP, 40 μl of 5 mM ddCTP, 50 μl of TE (50:1) buffer and 352.8 μl of sterile double-distilled water to give a final volume of 500 μl. Aliquot (18 μl for six reactions) and store at –70°C.

***Bst* 'short' termination 'G' mix:** Add 16.4 μl of 5 mM dATP to 16.4 μl of 5 mM dCTP, 8 μl of 0.5 mM c⁷dGTP, 16.4 μl of 5 mM dTTP, 54 μl of 5 mM ddGTP, 50 μl of TE (50:1) buffer and 338.8 μl of sterile double-distilled water to give a final volume of 500 μl. Aliquot (18 μl for six reactions) and store at −70°C.

***Bst* 'short' termination 'T' mix:** Add together 16.4 μl of 5 mM dATP, 16.4 μl of 5 mM dCTP, 16.4 μl of 5 mM c⁷dGTP, 8 μl of 0.5 mM dTTP, 60 μl of 5 mM ddTTP, 50 μl of TE (50:1) buffer and 332.8 μl of sterile double-distilled water to give a final volume of 500 μl. Aliquot (18 μl for six reactions) and store at −70°C.

**100 mM Calcium chloride:** 1.48 g of $CaCl_2$-$2H_2O$, make up to 100 ml with double-distilled water. Autoclave to sterilize. Store at 4°C.

**50 mM Calcium chloride:** 0.74 g of $CaCl_2$-$2H_2O$, make up to 100 ml with double-distilled water. Autoclave to sterilize. Store at 4°C.

**De-ionized formamide:** Stir formamide (Schwarz/Mann Biotech 800686) with amberlite MB resin, 10 g per 100 ml, for 1 h to de-ionize; filter through Whatman 3MM paper, store in a dark bottle at room temperature or 4°C.

**10× Denaturing buffer:** Add 2 ml of 1 M Tris–HCl, pH 9.5, 20 μl of 0.5 M EDTA, pH 8.0, 1 ml of 100 mM spermidine and make up to 10 ml with double-distilled water. Aliquot and store at −20°C.

**Diatomaceous earth (100 mg/ml):** Suspend 10 g of diatomaceous earth (Sigma D-5384) in 100 ml of distilled water in a 100 ml graduated cylinder, and let it settle down for 3 h. Decant the supernatant, and resuspend the pellet in 100 ml of 6 M guanidine-HCl, pH 6.4, 50 mM Tris–HCl, 20 mM EDTA.

**Diatomaceous earth wash buffer:** Add together 10 ml of 1 M Tris–HCl, pH 8.0, 2 ml of 0.5 M EDTA, pH 8.0, 500 ml of 100% ethanol and add double-distilled water to make 1 l.

**DNase-free RNase A:** Add 200 mg of RNase A (Sigma R-5500), 3.3 μl of 3 M sodium acetate, pH 4.5 and double-distilled water to make 10 ml. Boil for 10 min, aliquot and store at −20°C.

**1 M Dithiothreitol (DTT, Cleland's reagent):** 1.54 g of DTT (Calbiochem 233155) and double-distilled water to make 10 ml. Aliquot and store at $-20°C$.

**0.1 M DTT:** 154 mg of DTT in 10 ml of distilled water. Store at $-20°C$.

**Dye/formamide/EDTA solution:** 0.3% xylene cyanol FF, 0.3% bromophenol blue, 10 mM EDTA in de-ionized formamide.

**0.5 M EDTA, pH 8.0:** Dissolve 186.1 g of $Na_2$EDTA in approximately 400 ml of double-distilled water, adjust pH to 8.0 with 10 M NaOH, and adjust to 1 l final volume with distilled water.

**100 mM EDTA, pH 8.0:** 20 ml of 0.5 M $Na_2$EDTA, pH 8.0 and 80 ml of double-distilled water to make 100 ml.

**95% Ethanol/0.12 M Sodium acetate (ethanol/acetate):** 95 ml of 100% ethanol, 4 ml of 3 M sodium acetate, pH 4.5 and 1 ml of double-distilled water to give a final volume of 100 ml.

**5 mg/ml Ethidium bromide (EtBr):** Add 500 mg of EtBr (Sigma E-8751) to double-distilled water to make 100 ml.

**FE (formamide/EDTA):** Make fresh using 10 µl of double-distilled water, 10 µl of 100 mM EDTA, pH 8.0 and 100 µl of de-ionized formamide.

**FE (formamide/EDTA) with Blue Dextran:** Make fresh using 10 µl of double-distilled water, 10 µl of 100 mM EDTA, pH 8.0 containing blue dextran and 100 µl of de-ionized formamide.

**10× Fill-in/kinase buffer:** 5 ml of 1 M Tris–HCl, pH 7.6, 1 ml of 1 M $MgCl_2$, 100 µl of 1 M DTT, 500 µl of 1 mg/ml BSA and 3.4 ml of double-distilled water to give a final volume of 10 ml.

**Fill-in deoxynucleotide preparation:** To make 4 ml of the fill-in nucleotides at a concentration of 0.25 mM of each nucleotide, combine 500 µl of PCR dNTPs (2 mM) and 3500 µl of double-distilled water. Aliquot this into 0.5 ml Eppendorf tubes with 10 µl in each tube. To make 4 ml of these nucleotides at a final concentration of 0.25 mM from the stock 100 mM solutions, add 10 μl of 100 mM dATP, 10 μl of 100 mM dCTP, 10 μl of 100 mM dGTP, 10 μl of 100 mM dTTPand 3.6 ml of double-distilled water. Aliquot into 0.5 microcentrifuge tubes with 10 μl in each tube.

**FSB (frozen storage buffer for preparation of competent cells:** 50 ml of sterile 1 M KOAc, pH 7.0, 3.73 g KCl, 4.45 g MnCl$_2$.4H$_2$O, 0.74 g CaCl$_2$.2HCL, 0.4 g hexaminyl-CoCl$_2$, 50 ml ultrapure glycerol, distilled water to 500 ml final volume after adjusting pH to 6.4 with 0.1 M HCl. Sterilize by filtering through a 0.22 μm membrane and store aliquoted at 4°C or −20°C.

**GET/lysozyme solution:** 50 mM glucose, 25 mM Tris–HCl, pH 8.0, and 10 mM EDTA, pH 8.0 in double-distilled water. 0.9 g of D-glucose, 2.5 ml of 1.0 M Tris–HCl, pH 8.0, 2 ml of 0.5 M EDTA, pH 8.0 and double-distilled water to 100 ml. Filter sterilize and store at 4°C. Add 2 mg/ml lysozyme (Sigma L-6876) just before use.

**20% Glucose:** 20 g of D-glucose and double-distilled water to 100 ml. Filter sterilize.

**6 M Guanidine-HCl, pH 6.4, 50 mM Tris–HCl, 20 mM EDTA:** 573.18 g of guanidine-HCl (Sigma G-4505), 50 ml of 1 M Tris–HCl, pH 7.6, 40 ml of 0.5 M EDTA, pH 8.0 and double-distilled water to 1 l.

**5× *Hind*/DTT buffer:** Mix equal volumes of 10× *Hin*d with 0.1 M DTT. Store at 4°C.

**10× *Hin*d buffer:** Mix 0.5 ml of 2 M Tris–HCl, pH 7.6, 0.7 ml of 1 M MgCl$_2$, 0.35 g of NaCl and distilled water to give 10 ml. Store at −20°C.

**1 M HEPES, pH 7.5:** 23.83 g of HEPES (Sigma H-3375) and double-distilled water to make 100 ml. Adjust pH to 7.5 with KOH (store at 4°C).

**1 M Isocitrate (sodium salt-dihydrate):** Add 29.41 g of tri-sodium isocitrate-2H$_2$O (Sigma C-7254) and double-distilled water to 100 ml.

**Isopropyl β-D-thiogalactopyranoside (IPTG):** Add 250 mg of IPTG (Sigma I-5502) to double-distilled water to make 10 ml. Aliquot and store at −20°C.

**Kanamycin sulfate (Kan):** Stock of 5 mg/ml in sterile double-distilled water. 0.5 g of kanamycin (Boehringer Mannheim 106 801) and sterile double-distilled water to 100 ml. Add to media for final concentration of 20 µg/ml.

**1 M KCl (potassium chloride):** 7.5 g of KCl and double-distilled water to 100 ml.

**10× Kinase buffer:** Add together 5 ml of 1 M Tris–HCl, pH 7.6, 1 ml of 1 M $MgCl_2$, 1 ml of 1 M DTT and make up to 10 ml with sterile double-distilled water. Store in 25 ml aliquots at −20°C.

**Klenow dilution buffer (KDB):** 50 mM Tris–HCl pH 7.5, 0.1 M $(NH_4)_2SO_4$, 1 mM DTT or 10 mM 2-mercaptoethanol, 1 mg/ml BSA.

**Klenow labeling mix:** 7.5 µM each of dCTP, 7-deaza-dGTP and dTTP.

**Klenow termination mixes:** all mixes contain either 240 µM ddATP, 100 µM ddCTP, 120 µM ddGTP or 400 µM ddTTP, the corresponding dNTP at 25 µM and the other three dNTPs at 250 µM. Aliquot in small amounts and store at −20°C.

**Lambda plates:** 10 g of bacto-tryptone (Difco 0123-01-1), 15 g of bacto-agar (Difco 0140-01), 2.5 g of NaCl and make up to 1 l with double-distilled water. Autoclave to sterilize and pour into sterile Petri dishes.

**Lambda top agar:** 10 g of bacto-tryptone (Difco 0123-01-1), 10 g of bacto-agar (Difco 0140-01), 5 g of NaCl and make up to 1 l with double-distilled water. Autoclave to sterilize.

**LB medium:** 10 g of bacto-tryptone (Difco 0123-01-1), 5 g of bacto-yeast extract (Difco 0127-05-3), 10 g of NaCl and double-distilled water to 1 l. Autoclave to sterilize.

**LB plates:** 10 g of bacto-tryptone (Difco 0123-01-1), 5 g of bacto-yeast extract (Difco 0127-05-3), 10 g of NaCl, 15 g of bacto agar (Difco 0140-01) and make up to 1 l with double-distilled water. Autoclave to sterilize, cool to 55°C, add antibiotic if desired, and pour into sterile Petri dishes.

**10× Ligation buffer:** Add 5 ml 1 M Tris–HCl, pH 7.6 to 1 ml of 1 M $MgCl_2$ 1 ml of 1 M DTT, 1 ml of 100 mM rATP, 2.5 mg of BSA and sterile double-distilled water to give 10 ml. Store in 25 ml aliquots at −20°C.

**Loading dye:** Add together 0.03 g of xylene cyanole FF, 0.03 g of bromophenol blue, 0.2 ml of 0.5 M EDTA, pH 8.0 and deionized formamide to make 10 ml.

**Lysozyme solution:** 5 ml of 1 M Tris–HCl, pH 8.0, 2 ml of 0.5 M EDTA, 0.5 g of lysozyme (Sigma L-6876) and sterile double-distilled water to give 100 ml. Make fresh.

**3 M KOAc (potassium acetate), pH 4.5:** 294.45 g of potassium acetate. Dissolve in 300 ml of double-distilled water, adjust pH with glacial acetic acid, bring to a final volume of 1 l with double-distilled water.

**M-9 agar:** Add 15 g of agar to 870 ml of double-distilled water in a 2 l Erlenmeyer flask and autoclave. Also autoclave a 100 ml graduated cylinder capped with aluminum foil to use for measuring the sterile M-9 salts later. Swirl the agar gently and carefully upon removal from the autoclave to disperse any undissolved agar. Allow to cool in a 55°C water bath. When at 55°C, add the ingredients called for in the M-9 liquid medium recipe, omitting the water. Be sure to use sterile pipettes or graduated cylinders, as this mixture cannot be autoclaved. Immediately pour into sterile Petri dishes, using a sterile procedure.

**M-9 Medium (liquid):** 100 ml of 10× M-9 salts, 1 ml of 1 M $MgSO_4$ (autoclaved), 10 ml of 20% glucose (filter sterilized), 1 ml of 1% thiamine (filter sterilized), 10 ml of 100 mM $CaCl_2$ (autoclaved) and sterile double-distilled water to give a final volume of 1 l.

**10× M-9 Salts:** 60 g of $Na_2HPO_4$ (sodium phosphate, dibasic), 30 g of $KH_2PO_4$ (potassium phosphate, monobasic), 5 g of NaCl, 10 g of $NH_4Cl$ and add double-distilled water to give 1 l. Autoclave.

**1 M $MgCl_2$:** 20.33 g of $MgCl_2$-$6H_2O$ and make up to 100 ml with double-distilled water.

**1 M $MgSO_4$:** 12.04 g of $MgSO_4$ and double-distilled water to give a final volume of 100 ml. Autoclave.

**1 M $MnCl_2$:** 1.98 g of $MnCl_2$ (Sigma M-8530) and make up to 10 ml with double-distilled water. Store protected from light.

**10× $Mn^{2+}$/isocitrate buffer:** 50 µl of 1 M $MnCl_2$, 150 µl of 1 M isocitrate, 250 µl of glycerol and 550 µl of double-distilled water to give a final volume of 1 ml.

**1 M MOPS, pH 7.5**: 20.93 g of MOPS (Sigma M-1254). Dissolve in 80 ml of double distilled water, adjust pH to 7.5 with 1 M NaOH, and adjust volume to 100 ml.

**10× MOPS buffer:** Add 400 µl of 1 M MOPS, pH 7.5, 170 µl of 3 M NaCl, 100 µl of 1 M MgCl$_2$ and 330 µl of double-distilled water to give a final volume of 1 ml.

**2.7 M MOPS (acid form):** 5.65 g of MOPS (acid form) and double-distilled water to 10 ml.

**MOPS-acid buffer:** Add together 500 µl of 2.7 M MOPS (acid form), 100 µl of 1 M MgCl$_2$ and 400 µl of double-distilled water to give a final volume of 1ml.

**10× MTBE (modified Tris-borate-EDTA buffer):** Add together 162 g of Tris base, 27.5 g of boric acid, 9.3 g of EDTA and double-distilled water to 1 l.

**3 M NaCl (sodium chloride)**: 17.53 g of NaCl and double-distilled water to make 100 ml.

**10 M NaOH:** 400 g of NaOH and make up to 100 ml with double-distilled water.

**1 M NaOH:** 10 ml of 10 M NaOH and double-distilled water to give 100 ml.

**10× PCR buffer:** Add 5 ml of 1 M KCl to 1 ml of 1 M Tris–HCl, pH 8.5, 150 µl of 1 M MgCl$_2$ and make up to 10 ml with double-distilled water.

**PCR deoxynucleotide preparation:** To make 12.5 ml of the PCR nucleotides at a concentration of 2 mM each nucleotide, combine the following: 250 µl of 100 mM dATP, 250 µl of 100 mM dCTP, 250 µl of 100 mM dGTP, 250 µl of 100 mM dTTP and 11.5 ml of double-distilled water. Aliquot and store at −20°C.

**20% PEG/2.5 M NaCl:** 7.3 g of NaCl, 10 g of PEG (mol. wt 8000) (Fisher P156-3). Dissolve in 30 ml of double-distilled water by stirring, and then adjust the volume to 50 ml.

**50% PEG/0.5 M NaCl:** 5.85 g of NaCl, 100 g of PEG (mol. wt 8000) (Fisher P156-3). Dissolve in 100 ml of double-distilled water by stirring, and then adjust the volume to 200 ml.

**PEG:TE rinse solution:** Combine 250 µl of 1 M Tris–HCl, pH 8.0, 50 µl of 0.5 M EDTA, pH 8.0, 12.5 ml of 20% PEG/2.5 M NaCl and double-distilled water to a volume of 37.5 ml

**Phenol, TE-saturated:** add an equal volume of 10 mM Tris–HCl, pH 7.5–8.0, 1 mM EDTA, pH 8.0 to ultrapure phenol, mix well, allow phases to separate, remove and discard upper (aqueous) phase. Repeat until the pH of the aqueous phase is between 7.5 and 8.0. Store at 4°C

**Phenol:chloroform:isoamyl alcohol (25:25:1):** 100 ml of TE-saturated phenol, 100 ml of chloroform, 4 ml of isoamyl alcohol to give a total volume of 204 ml.

**Restriction enzyme assay buffer, 10× low salt:** 1 ml of 1 M Tris–HCl, pH 7.6, 1 ml of 1.0 M $MgCl_2$, 0.1 ml of 1.0 M DTT and double-distilled water to give a volume of 10 ml.

**Restriction enzyme assay buffer, 10× medium salt:** Combine 1.7 ml of 3 M NaCl, 1 ml of 1.0 M Tris–HCl, pH 7.6, 1 ml of 1.0 M $MgCl_2$, 0.1 ml of 1 M DTT and make up to 10 ml with double-distilled water.

**Restriction enzyme assay buffer, 10× high salt:** Add 3.3 ml of 3 M NaCl to 5 ml of 1 M Tris–HCl, pH 7.6, 1 ml of 1 M $MgCl_2$, 0.1 ml of 1 M DTT and double-distilled water to give a final volume of 10 ml.

**Restriction enzyme assay buffer, 10× *Sma*I:** 2 ml of 1 M KCl, 1 ml of 1 M Tris–HCl, pH 7.6, 1 ml of 1 M $MgCl_2$, 0.1 ml of 1 M DTT and double-distilled water to make 10 ml.

**RNase T1:** 100 µl of RNase T1 (Sigma R-8251) (100 000 U/0.2 ml), 25 µl of 1 M Tris–HCl, pH 7.6 and 375 µl of double-distilled water to give a final volume of 500 µl.

**Silanizing reagent:** 5% solution of dichlorodimethyl silane in 1,1,1-trichloroethane.

**2 M Sodium acetate:** 27.22 g of sodium acetate-3$H_2$O and double-distilled water to give a final volume of 100 ml.

**3 M Sodium acetate, pH 4.5:** 408.24 g of sodium acetate-3$H_2$O. Dissolve in approximately 300 ml of double-distilled water, adjust pH to 4.8 with glacial acetic acid and bring to a final volume of 1 l with double-distilled water.

**20× SSC (standard saline-citrate):** 17.53 g of NaCl and 8.82 g of sodium citrate. Dissolve in approximately 80 ml of double-distilled water, adjust pH to 7.0 with hydrochloric acid and bring the final volume to 100 ml.

**1× STB buffer:** 25 g of sucrose, 5 ml of 1 M Tris–HCl, pH 8.0 and double-distilled water to give a volume of 100 ml. Filter sterilize and store at 4°C.

**5× T7 labeling mix:** 7.5 mM each dCTP, dGTP and dTTP, 0.1 M DTT: 154 g of dithiothreitol in 10 ml of distilled water. Store at −20°C.

**10× T7 sequencing buffer:** 200 mM Tris–HCl, pH 7.5, 50 mM $MgCl_2$, 250 mM NaCl.

**T7 termination mixes:** All mixes are 80 μM each of dATP, dCTP, dGTP and dTTP, 50 mM NaCl and 8 μM of the appropriate ddNTP. Aliquot in small amounts and store at −20°C.

**20× TAE buffer:** Add 96.9 g of Tris base to 32.8 g of sodium acetate-$3H_2O$ and 14.9 g of EDTA. Dissolve in approximately 700 ml of double-distilled water, adjust the pH to 8.3 with glacial acetic acid, and bring to 1 l with double-distilled water.

**5× *Taq* dilution buffer:** Add together 16 ml of 1 M Tris–HCl, pH 9.0, 4 ml of 1 M $(NH_4)_2SO_4$, pH 9.0, 1 ml of 1 M $MgCl_2$ and 19 ml of double-distilled water to give a final volume of 40 ml.

**5× *Taq* reaction buffer:** Combine 16 ml of 1 M Tris–HCl, pH 9.0, 4 ml of 1 M $(NH_4)_2SO_4$, pH 9.0, 1 ml of 1 M $MgCl_2$ 2 ml of DMSO and 17 ml of double-distilled water to give a final volume of 40 ml.

**5× *Taq* 'A' termination mix:** Add 20 μl of 20 mM dATP to 80 μl of 20 mM dCTP, 240 μl of 10 mM $c^7dGTP$, 80 μl of 20 mM dTTP, 1920 μl of 5 mM ddATP, 640 μl of TE (50:1) buffer and 3420 μl of sterile double-distilled water to give a final volume of 6.4 ml

**5× *Taq* 'C' termination mix:** Add together 80 μl of 20 mM dATP, 20 μl of 20 mM dCTP, 240 μl of 10 mM $c^7dGTP$, 80 μl of 20 mM dTTP, 960 μl of 5 mM ddCTP, 640 μl of TE (50:1) buffer and 4380 μl of sterile double-distilled water to give a final volume of 6.4 ml

**5× *Taq* 'G' termination mix:** Add 160 μl of 20 mM dATP to 160 μl of 20 mM dCTP, 120 μl of 10 mM c$^7$dGTP, 160 μl of 20 mM dTTP, 320 μl of 5 mM ddGTP, 1280 μl of TE (50:1) buffer and 10 600 μl of sterile double-distilled water to give a final volume of 12.8 ml

**5× *Taq* 'T' termination mix:** Combine 160 μl of 20 mM dATP, 160 μl of 20 mM dCTP, 480 μl of 10 mM c$^7$dGTP, 40 μl of 20 mM dTTP, 3200 μl of 5 mM ddTTP, 1280 μl of TE (50:1) buffer and 7480 μl of sterile double-distilled water to give a final volume of 12.8 ml

**10× TB Salts:** Combine 2.31 g of $KH_2PO_4$, 12.54 g of $K_2HPO_4$ (potassium phosphate, dibasic) and double-distilled water to 100 ml. Autoclave.

**TEMED (*N,N,N′,N′*-tetramethylethylenediamine):** Kodak T-7024. Store protected from light at 15°C.

**Terrific Broth (TB):** 12 g of bacto-tryptone, 24 g of yeast extract, 4 ml of glycerol and double-distilled water to give a volume of 900 ml. Autoclave, cool and add 100 ml of 10× TB salts. Adjust the final volume to 1 l with sterile double-distilled water.

**TE (10:0.1) buffer:** Combine 10 ml of 1 M Tris–HCl, pH 7.6, 0.2 ml of 0.5 M EDTA, pH 8.0 and double-distilled water to make 1 l.

**TE (10:1) buffer:** Add 10 ml of 1 M Tris–HCl, pH 7.6, 2 ml of 0.5 M EDTA, pH 8.0 and double-distilled water to give a volume of 1 l.

**TE (100:10) buffer:** Combine 100 ml of 1 M Tris–HCl, pH 7.6, 20 ml of 0.5 M EDTA and make up to a final volume of 1 l with double-distilled water.

**TE (50:1) buffer:** Add together 0.5 ml of 1 M Tris–HCl, pH 7.6, 0.1 ml of 100 mM EDTA, pH 8.0 and 9.4 ml of double-distilled water to give a final volume of 10 ml.

**TE-RNase solution:** 0.6 ml of 1 M Tris–HCl, pH 7.6, 240 μl of 0.5 M EDTA, pH 8.0, 25 μl of 20 mg/ml RNase A (Sigma R-5500) and 11.2 ml of double-distilled water to give a final volume of 12.0 ml.

**Tetracycline stock (Tet):** Stock of 10 mg/ml in 50% ethanol in sterile double-distilled water. 1 g of tetracycline (Sigma T-3383), 50 ml of 100% ethanol and sterile double-distilled water to 100 ml. Store at 4°C in the absence of light. Add to media for final concentration of 20 μg/ml.

**1% Thiamine:** Combine 100 mg of thiamine (Sigma T-4625) and sterile double-distilled water to give 10 ml (filter sterilized).

**10× TM buffer:** 5 ml of 1.0 M Tris–HCl, pH 8.0, 1.5 ml of 1.0 M $MgCl_2$ and add sterile double-distilled water to give a final volume of 10 ml.

**1 M Tris–HCl, pH 7.6, 8.0, 8.5, 9.0, 9.5:** 121.1 g of Tris base and double-distilled water to 800 ml. Adjust pH with concentrated HCl and then add double-distilled water to 1 l.

**50:2:10 TTE:** 5 ml of 1 M Tris–HCl, pH 8.0, 2 ml of 0.5 M EDTA, pH 8.0, 2 ml of Triton X-100 (Sigma X-100) and double-distilled water to 100 ml.

**TTE:** 500 μl of 1 M Tris–HCl, pH 8.0, 250 μl of Triton X-100 (Sigma X-100), 10 μl of 0.5 M EDTA, pH 8.0 and double-distilled water to 50 ml.

**2× TY medium:** 16 g of bacto-tryptone (Difco 0123-01-1), 10 g of bacto-yeast extract (Difco 0127-05-3), 5 g of NaCl and double-distilled water to 1 l. Autoclave.

**X-Gal (5-bromo-4-chloro-3-indolyl β-D-galactopyranoside):** 200 mg of X-gal (Sigma B-4252) and dimethylformamide (DMF) to 10 ml. Aliquot and store protected from light at −20°C.

## PRIMERS

ABI Forward primer sequence: 5′ TGT AAA ACG ACG GCC AGT 3′
ABI Forward Aminolink-primer sequence: 5′ 5TG TAA AAC GAC GGC CAG T 3′
ABI Reverse primer sequence: 5′ CAG GAA ACA GCT ATG ACC 3′
ABI Reverse Aminolink-primer sequence: 5′ 5CA GGA AAC AGC TAT GAC C 3′

# *TAQ* CYCLE SEQUENCING REAGENT PREPARATION

**5× *Taq* reaction buffer** (for preparation, see p. 149). This buffer will be added separately with the A, C, G and T nucleotide mixes for ease in reaction pipetting. One 40 ml preparation of buffer will be sufficient for one batch (~200 tubes) of A, C, G and T mix aliquots.

***Taq* dilution buffer** (for preparation, see p. 149) This is routinely distributed into 30 µl aliquots in 0.5 ml microcentrifuge tubes (~ 200 per batch).

**TE (50:1) buffer** (for preparation, see p. 150)

**Fluorescently labeled primers.** The ratio of the $A_{260}$ to $A_x$ for an pure 18-mer-dye labeled primer should be approximately 3:1 (ABI User Bulletin # 11, p. 13). For different length base primers the ratio is given by:

$$A_{260}(n/18) / A\text{x} \qquad \text{Where } n = \text{length of primer in bases}$$

Prepare a 100× stock solution (40 µM); an example calculation for a dry tube of an 18-mer with an OD of 1.00 is shown below (remembering that JOE is the dye-labeled primer for the A reaction, FAM is for C, TAMRA is for G, and ROX is for T):

$$1.00 \text{ OD}(37 \text{ µg/OD})(\text{mol} \times \text{mer}/320 \text{ g})(10^{12} \text{ pmol/mol})(g/10^{6} \text{ µg})(1/18\text{-mer})(\text{µl}/40 \text{ pmol}) = x \text{ µl}$$

In this example, x = 160 µl, and 160 µl of double-distilled water should be added to the dried tube of fluorescent primer for a concentration of 40 µM (40 pmol/µl). From this 100× stock of 40 µM make 1:100 dilutions. To make the amount of primer aliquoted the same as the amount of mixes per batch, dilute either fluorescent forward or reverse primers as follows: 64 µl of 40 µM A or C primer diluted with 6.3 ml of double-distilled water to give 6.4 ml, and 128 µl of 40 µM G or T primer diluted with 12.7 ml of double-distilled water to give 12.8 ml.

For the A and C primers, distribute the 1× (0.4 µM solution) into 30 µl aliquots, and for the G and T primers, distribute them into 60 µl aliquots. The primers aliquots are stored in clear 0.5 ml microcentrifuge tubes which are labeled with blue, green, red or yellow markers for A, C, G, or T primers, respectively. (Note: the current primers work optimally at the effective concentration of 0.4 µM, however with each new fluorescent primer preparation, the optimal concentration must be determined.) The primers should be stored at −20 or −70°C.

### 5× *Taq* cycle sequencing mixes

Working dilutions of 20 mM are made for dATP, dCTP and dTTP based on using one complete tube of 20 mM stock per batch of mixes. c$^7$dGTP is purchased at a concentration of 10 mM (1080 µl are needed for one batch each of A, C, G and T mixes, so slightly more than five tubes will be needed – each tube contains 200 µl).

| 20 mM dATP | 20 mM dCTP | 20 mM dTTP |
|---|---|---|
| 95 µl 100 mM dATP | 95 µl 100 mM dCTP | 80 µl 100 mM dTTP |
| 47.5 µl TE (50:1) | 47.5 µl TE (50:1) | 40 µl TE (50:1) |
| 332.5 µl double-distilled water | 332.5 µl double-distilled water | 280 µl double-distilled water |
| 475 µl | 475 µl | 400 µl |

The concentration of deoxy and dideoxy nucleotides in the mixes are shown below, followed by the recipe for one 200 tube batch of each of the four mixes.

| | A | C | G | T |
|---|---|---|---|---|
| dATP | 62.5 µM | 250 µM | 250 µM | 250 µM |
| dCTP | 250 µM | 62.5 µM | 250 µM | 250 µM |
| c$^7$dGTP | 375 µM | 375 µM | 94 µM | 375 µM |
| dTTP | 250 µM | 250 µM | 250 µM | 62.5 µM |
| ddATP | 1.5 mM | – | – | – |
| ddCTP | – | 0.75 mM | – | – |
| ddGTP | – | – | 0.125 mM | – |
| ddTTP | – | – | – | 1.25 mM |

*Appendix A: Solution recipes*

For one batch (200 tubes) of each nucleotide mix:

|  | A (µl) | C (µl) | G (µl) | T (µl) |
|---|---|---|---|---|
| 20 mM dATP | 20 | 80 | 160 | 160 |
| 20 mM dCTP | 80 | 20 | 160 | 160 |
| 10 mM c$^7$dGTP | 240 | 240 | 120 | 480 |
| 20 mM dTTP | 80 | 80 | 160 | 40 |
| 5 mM ddATP | 1920 | – | – | – |
| 5 mM ddCTP | – | 960 | – | – |
| 5 mM ddGTP | – | – | 320 | – |
| 5 mM ddTTP | – | – | – | 3200 |
| TE (50:1) | 640 | 640 | 1280 | 1280 |
| Double-distilled water | 3420 | 4380 | 10 600 | 7480 |
| Total | 6400 | 6400 | 12 800 | 12 800 |

To each of these mix solutions, an equal volume of 5× *Taq* reaction buffer is added, so 6.4 ml is added to A and C, and 12.8 ml is added to G and T. This mix/buffer solution is distributed into 0.5 ml microcentrifuge tubes in 60 or 120 µl aliquots (60 for A and C/120 for G and T). The simplest way to distribute the 60 µl aliquots is 2× 30 µl using the Eppendorf repeat pipette set on 3 with the 0.5 ml Combitips, and for the 120 µl aliquots use 1× 100 µl with the 5 ml Combitip plus 1× 20 µl with the 0.5 ml Combitip. The mixes should be stored at −20 or −70°C.

# OLIGONUCLEOTIDE UNIVERSAL PRIMERS USED FOR DNA SEQUENCING

M13 (–21) universal forward: 5′-TGT-AAA-ACG-ACG-GCC-AGT-3′.
M13 (–40) universal forward: 5′-GTT-TTC-CCA-GTC-ACG-AC-3′.
M13/pUC reverse primer: 5′-CAG-GAA-ACA-GCT-ATG-ACC-3′.
T7 primer: 5′-TAA-TAC-GAC-TCA-CTA-TAG-GG-3′.
SP6 primer: 5′-ATT-TAG-GTG-ACA-CTA-TAG-3′.

# APPENDIX B: COMMONLY USED BACTERIAL STRAINS

C600      *F⁻, e14, mcrA, thr-1 supE44, thi-1, leuB6, lacY1, tonA21, l–*. For λ lambda (gt10) libraries, grows well in L broth, 2× TY, plate on NZYDT + Mg.

DH1      *F⁻, recA1, endA1, gyrA96, thi-1, hsdR17 (rk⁻, mk⁺), supE44, relA1, l⁻*. For plasmid transformation, grows well on L broth and plates.

XL1Blue-MRF′      *D(mcrA)182, D(mcrCB-hsdSMR-mrr)172,endA1, supE44, thi-1, recA, gyrA96, relA1, lac, l⁻, [F′proAB, lac IqZDM15, Tn10 (tetʳ)]*. For plating or glycerol stocks, grow in LB with 20 μg/ml of tetracycline. For transfection, grow in tryptone broth containing 10 mM MgSO₄ and 0.2% maltose (no antibiotic – Mg²⁺ interferes with tetracycline action). For picking plaques, grow glycerol stock in LB to an OD of 0.5 at 600 nm (2.5 h). When at 0.5, add MgSO₄ to a final concentration of 10 mM.

SURE cells (Stratagene)      *e14(mcrA), D(mcrCB- hsdSMR-mrr)171, sbcC, recB, recJ, umuC::Tn5 (kanʳ), uvrC, supE44, lac, gyrA96, relA1, thi-1, end A1[F′proAB, lacIqDM15, Tn10(tetʳ)]*. An uncharacterized mutation enhances the α-complementation to give a more intense blue color on plates containing X-gal and IPTG.

GM272      *F⁻, hsdR544 (rk⁻, mk⁻), supE44, supF58, lacY1 or ΔlacIZY6, galK2, galT22, metB1m, trpR55, l⁻*. For plasmid transformation, grows well in 2× TY, TYE, L broth and plates.

HB101      *F⁻, hsdS20 (rb⁻, mb⁻), supE44, ara14, galK2, lacY1, proA2, rpsL20 (strᴿ), xyl-5, mtl-1, l⁻, recA13, mcrA(+), mcrB(−)*. For plasmid transformation, grows well in 2× TY, TYE, L broth and plates.

JM101      *supE, thi, Δ(lac-proAB), [F′, traD36, proAB, lacIqZΔM15], restriction: (rk⁺, mk⁺), mcrA+*. For M13 transformation, grow on minimal medium to maintain F episome, grows well in 2× TY, plate on TY or lambda agar.

XL-1 blue      *recA1, endA1, gyrA96, thi, hsdR17 (rk⁺, mk⁺), supE44, relA1, l⁻, lac, [F′, proAB, lacIqZΔM15,*

Tn*10 (tet<sup>R</sup>)].* For M13 and plasmid transformation, grow in 2× TY + 10 µg/ml Tet, plate on TY agar + 10 µg/ml Tet (Tet maintains F episome).

GM2929     From B. Bachman, Yale E.coli Genetic Stock Center; M.Marinus strain; sex F⁻; (*ara-14, leuB6, fhuA13, lacY1, tsx-78, supE44, [glnV44], galK2, galT22, l⁻, mcrA, dcm-6, hisG4,[Oc], rfbD1, rpsL136, dam-13::*Tn*9, xyl-5, mtl-1, recF143, thi-1, mcrB, hsdR2*).

MC1000     (*araD139, D[ara-leu]7679, galU, galK, D[lac]174, rpsL, thi-1*). Obtained from the McCarthy laboratory at the University of Oklahoma.

ED8767     F⁻, *e14-[mcrA], supE44, supF58, hsdS3[r_B.m_B.], recA56, galK2, galT22, metB1, lac-3 or lac3Y1⁻* Obtained from Nora Heisterkamp and used as the host for *abl* and *bcr* cosmids.

# APPENDIX C: SUPPLIERS

| Abbreviation | Company name | Abbreviation | Company name |
|---|---|---|---|
| AIP | Amersham International plc | GRI | Genetic Research Instrumentation Ltd |
| APP | Appligene | HSI | Hoefer Scientific Instruments |
| BDH | BDH Laboratories | IBI | International Biotechnologies, Inc. |
| | | IPI | IPI Medical Products |
| BHM | Boehringer Mannheim | NBL | Northumbria Biologicals Ltd |
| BRL | BioRad Laboratories Ltd | PMB | Pharmacia Biosystems |
| EKL | Eastman Kodak Ltd | PML | Promega Ltd |
| FIL | Flowgen Instruments Ltd | SCC | Sigma Chemical Corporation |
| GBL | Gibco–BRL | STG | Stratagene |

**Amersham International plc.,** Amersham Place, Little Chalfont, Bucks HP7 9NA, UK. Tel (0800) 616928, (01494) 54400. Fax (0800) 616927, (01494) 54266, 26236. South Clearbrook Drive, Arlington Heights, IL 60005, USA. Tel (708) 593 6300. Fax (708) 593 8010.

**Anachem Ltd,** Anachen House, 20 Charles Street, Luton, Beds LU2 0EB, UK. Tel (01582) 456666. Fax (01582) 391768.

**Applied Biosystems Ltd,** 7 Kingsland Grange, Woolston, Warrington, Cheshire WA1 4SR, UK. Tel 01925 825650. Fax 01925 28270.

**Du Pont NEN (UK) Ltd,** Wedgewood Way, Stevenage, Herts SG1 4QN, UK. Tel (01438) 734027. Fax (01438) 734379.

**Fisher Scientific Equipment,** Bishop Meadow Road, Loughborough, Leics LE11 0RG, UK. Tel (01509) 231166. Fax (01509) 231166.

**Flowgen Instruments Ltd,** Lynn Lane, Shenstone, Lichfield, Staffs WS14 0EE. Tel (01795) 429737. Fax (01795) 471185.

**Genetic Research Instrumentation Ltd,** Gene House, Dunmow Road, Felsted, Dunmow, Essex CM6 3LD, UK. Tel (01371) 821082. Fax (01371) 820131.

**Appligene,** Pinetree Centre, Durham Road, Birtely, Chester-le-Street, Co. Durham DH3 2TD, UK. Tel (0191) 4920022. Fax (0191) 4920617.
Parc d'Innovation, rue Geiler de Kaisersberg, BP 72, F-67402, Illkirch cedex, France. Tel 33 88 672267. Fax 33 88 671719 45.

**Beckman Instruments (UK) Ltd,** Oakley Court, Kingsmead Business Park, London Road, High Wycombe, Bucks HP11 IJU, UK. Tel (01494) 441181. Fax (01494) 463836.

**BioRad Laboratories Ltd,** BioRad House, Maylands Avenue, Hemel Hempstead. Herts HP2 7TD, UK. Tel (01442) 232522, (0800) 181134. Fax (01442) 259118.

**Boehringer Mannheim UK (Diagnostics and Biochemicals),** Bell Lane, Lewes, East Sussex BN7 1LG, UK. Tel (01273) 480444, (0800) 521578.
9115 Hague Road. PO Box 5044414, Indianapolis, IN 46250-0414, USA. Tel (800) 2621640. Fax (317) 576 2754.

**Calbiochem-Novabiochem (UK) Ltd,** Boulevard Industrial Park, Padge Road, Beeston, Nottingham NG9 2JR, UK. Tel (0800) 622935, (0115) 9430840. Fax (0115) 9430951.

**Clontech & Pharmingen Distributors**, Cambridge Biosciences, 24-25 Signet Court, Newmarket Road, Cambridge CB5 8LA, UK. Tel (01223) 316855. Fax (01223) 360732.

**ICN Biomedical Inc.,** 3300 Hyland Avenue, Costa Mesa, CA 92626, USA. Tel (800) 854 0530. Fax (800) 854 0530.
Unit 18, Thame Park Business Centre, Winman Road, Thame, Oxon OX9 3XA, UK. Tel (0800) 282474, (01844) 213366. Fax (0800) 614735, (01844) 213399.

**IPI Medical Products:** 7790 North Merrimac Niles, IL 60648, USA.

**Life Sciences Laboratories,** Sedgewick Road, Luton, Beds LU4 9DT, UK. Tel (01582) 597676. Fax (01582) 581495.

**Life Technologies,** 8451 Helgerman Court, Gaithersburg, MD 20884, USA. Tel (301) 8404152. Fax (301) 670 8539.

**Merk Ltd,** Merk House, Poole, Dorset BH15 4TD, UK. Tel (01202) 669700, (0800) 223344. Fax (01202) 665599.

**Millipore (UK) Ltd,** Blackmore Lane, Watford WD1 8YW, UK. Tel (01923) 816375. Fax (01923) 818297.

**NBL Gene Sciences Ltd,** South Nelson Road, Cramlington, Northumberland NE23 9WF, UK.Tel (01670) 732992. Fax (01670) 730454.

**New England Biolabs Ltd,** 67 Knowl Piece, Wilbury Way, Hitchin, Herts SG4 0TY, UK. Tel (01462) 420616. Fax (01462) 421057.
32 Tozer Road, Beverley, MA 01915-5599, USA. Tel (508) 927 5054. Fax (508) 921 1350.

**Perkin-Elmer Corporation,** 761 Main Avenue, Norwalk, CT 06859-33301, USA. Tel 203 762 1000. Fax 203 762 6000.

**Pharmacia Biotech,** 23 Grosvenor Road, St Albans, Herts AL1 3AW, UK. Tel (01727) 814000. Fax (01727) 814001. 800 Centennial Avenue, PO Box 1327, Pistcataway, NJ 08855-1327, USA. Tel. (201) 457 8000. Fax (201) 457 0557.

**Promega Corp.,** 20800 Woods Hollow Road, Madison, WI 53711-5399, USA. Tel (608) 274 4330. Fax (608) 277 2516.

**Promega Ltd,** Delta House, Chilworth Research Centre, Southampton SO16 7NS, UK. Tel (01703) 760225, (0800) 378994. Fax (01703) 767014, (0800) 181037.

**QIAGEN GmbH,** Max-Volmer Strasse 4, 40724 Hilden, Germany. Tel (02103) 892230. Fax (02103) 892222.

**QIAGEN Inc.,** 9600 De Soto Avenue, Chatsworth, CA 91311, USA. Tel (800) 426 8157. Fax (800) 718 2056.

**Schleicher & Schuell**, 10 Optical Avenue, PO Box 2021, Keene, NH 03431, USA. Tel (603) 352 3810. Fax (603) 357 3627.

**Schleicher & Schuell GmbH,** PO Box 4, D-37582 Dassel, Germany. Tel 5561 7910. Fax 5551 791536.

**Scientific Imaging Systems Ltd,** 36 Clifton Road, Cambridge CB1 4ZR, UK. Tel (01223) 242813. Fax (01223) 243036.

**Sigma Chemical Corp.,** Fancy Road, Poole, Dorset BH17 4QH. Tel (01202) 733114, (0800) 373731. Fax (01202) 715460. PO Box 14508, 3500 DeKalb Street, St Louis, MO 63178, USA. Tel (800) 848 7791. Fax (314) 771 5757.

**Stratagene Ltd,** 140 Cambridge Science Park, Milton Road, Cambridge CB4 4GF, UK. Tel (01223) 420955, (0800) 585370. Fax (01223) 420234. 11011 North Torrey Pines Road, La Jolla, CA 92037, USA. Tel (619) 535 5400, (800) 424 5444. Fax (619) 535 0034.

**United States Biochemical,** PO Box 22400, Cleveland, OH 44122, USA. Tel (216) 765 5000, (800) 321 9322. Fax (216) 464 5075, (800) 535 0898.

**Whatman International Ltd,** Whatman House, St Leonards Road, 20/20 Maidstone, Kent ME16 0LS, UK. Tel (01622) 676670. Fax (01622) 677011

**Worthington Biochemical Corporation,** Halls Mill Road, Freehold, NJ 07728, USA. Tel (908) 462 3838, (800) 445 9603. Fax (800) 368 3103, (908) 308 4453.

# INDEX